KITCHENER PUBLIC LIBRARY

D1290752

Fluid Mechanics
Demystified

Demystified Series

Fluid Mechanics Demystified

Merle C. Potter, Ph.D.

New York   Chicago   San Francisco   Lisbon   London
Madrid   Mexico City   Milan   New Delhi   San Juan
Seoul   Singapore   Sydney   Toronto

The McGraw·Hill Companies

Library of Congress Cataloging-in-Publication Data

Potter, Merle C.
    Fluid mechanics demystified / Merle C. Potter.
        p.   cm.—(Demystified)
    ISBN 978-0-07-162681-1 (alk. paper)
    1. Fluid mechanics. I. Title.
  TA357.P723   2009
  620.1′06—dc22   2009010124

McGraw-Hill books are available at special quantity discounts to use as premiums and sales promotions, or for use in corporate training programs. To contact a special sales representative, please visit the Contact Us page at www.mhprofessional.com.

Fluids Mechanics Demystified

Copyright © 2009 by The McGraw-Hill Companies, Inc. All rights reserved. Printed in the United States of America. Except as permitted under the United States Copyright Act of 1976, no part of this publication may be reproduced or distributed in any form or by any means, or stored in a data base or retrieval system, without the prior written permission of the publisher.

1 2 3 4 5 6 7 8 9 0 DOC/DOC 0 1 3 2 1 0 9 8

ISBN 978-0-07-162681-1
MHID 0-07-162681-6

Sponsoring Editor
  Judy Bass

Acquisitions Coordinator
  Michael Mulcahy

Editing Supervisor
  David E. Fogarty

Project Manager
  Preeti Longia Sinha, International
  Typesetting and Composition

Copy Editor
  Bhavna Gupta, International Typesetting
  and Composition

Proofreader
  Upendra Prasad, International
  Typesetting and Composition

Production Supervisor
  Richard C. Ruzycka

Composition
  International Typesetting and
  Composition

Art Director, Cover
  Jeff Weeks

Information contained in this work has been obtained by The McGraw-Hill Companies, Inc. ("McGraw-Hill") from sources believed to be reliable. However, neither McGraw-Hill nor its authors guarantee the accuracy or completeness of any information published herein, and neither McGraw-Hill nor its authors shall be responsible for any errors, omissions, or damages arising out of use of this information. This work is published with the understanding that McGraw-Hill and its authors are supplying information but are not attempting to render engineering or other professional services. If such services are required, the assistance of an appropriate professional should be sought.

ABOUT THE AUTHOR

Merle C. Potter, Ph.D., taught fluid mechanics for 42 years at Michigan Tech, The University of Michigan, and Michigan State University. He is the author of several engineering books, including *Thermodynamics Demystified*. Dr. Potter's research has been in theoretical and experimental fluid mechanics and energy analysis of buildings. He has received several teaching awards including the 2008 American Society of Mechanical Engineer's James Harry Potter Gold Medal.

CONTENTS

Contents

PREFACE

This book is intended to accompany a text used in that first course in fluid mechanics that is required in all mechanical engineering and civil engineering departments, as well as several other departments. It provides a succinct presentation of the material so that students more easily understand the difficult parts. Many fluid mechanics texts are very long and it is often difficult to ferret out the essentials due to the excessive verbiage. This book presents those essentials.

We have included a derivation of the Navier-Stokes equations with several solved flows. It is not necessary, however, to include them if the elemental approach is selected. Either method can be used to study laminar flow in pipes, channels, between rotating cylinders, and in laminar boundary layer flow.

The basic principles upon which a study of fluid mechanics is based are illustrated with numerous examples and practice exams that allow students to develop their problem-solving skills. The solutions to all problems in the practice exams are included at the end of the book. Examples and problems are presented using SI metric units.

The practice exams at the end of each chapter contain four-part, multiple-choice problems similar to those found on national exams, such as the Fundamentals of Engineering exam (the first of two exams required in the engineering registration process) or the Graduate Record Exam (required when applying for most graduate schools). There are also partial-credit practice exams at the end of each chapter. Engineering courses do not, in general, utilize multiple-choice exams but it is quite important that students gain experience in taking such exams. This book allows that experience, if desired. If one correctly answers 50% or more multiple-choice questions correctly, that is quite good.

The mathematics required is that of other engineering courses except that if the study of the Navier-Stokes equations is selected then partial differential equations are encountered. Some vector relations are used but not at a level beyond most engineering curricula.

If you have comments, suggestions, or corrections or simply want to opine, please email me at MerleCP@sbcglobal.net. It is impossible to write a book free of errors but if we're made aware of them, we can have them corrected in future printings. So, please send me an email when you discover one.

Merle C. Potter P.E., Ph.D.

Fluid Mechanics
Demystified

CHAPTER 1

The Essentials

Fluid mechanics is encountered in almost every area of our physical lives. Blood flows through our veins and arteries, a ship moves through water, airplanes fly in the air, air flows around wind machines, air is compressed in a compressor, steam flows around turbine blades, a dam holds back water, air is heated and cooled in our homes, and computers require air to cool components. All engineering disciplines require some expertise in the area of fluid mechanics.

In this book we will solve problems involving relatively simple geometries, such as flow through a pipe or a channel, and flow around spheres and cylinders. But first, we will begin by making calculations in fluids at rest, the subject of fluid statics.

The math required to solve the problems included in this book is primarily calculus, but some differential equations will be solved. The more complicated flows that usually are the result of more complicated geometries will not be presented.

In this first chapter, the basic information needed in our study will be presented.

1.1 Dimensions, Units, and Physical Quantities

Fluid mechanics is involved with physical quantities that have dimensions and units. The nine basic dimensions are mass, length, time, temperature, amount of a substance, electric current, luminous intensity, plane angle, and solid angle. All other quantities can be expressed in terms of these basic dimensions; for example, force can be expressed using Newton's second law as

$$F = ma \tag{1.1}$$

In terms of dimensions we can write (note that F is used both as a variable and as a dimension)

$$F = M \frac{L}{T^2} \tag{1.2}$$

where F, M, L, and T are the dimensions of force, mass, length, and time. We see that force can be written in terms of mass, length, and time. We could, of course, write

$$M = F \frac{T^2}{L} \tag{1.3}$$

Units are introduced into the above relationships if we observe that it takes 1 newton to accelerate 1 kilogram at 1 meter per second squared, i.e.,

$$N = kg \cdot m/s^2 \tag{1.4}$$

This relationship will be used often in our study of fluids. In the SI system, mass will always be expressed in kilograms, and force in newtons. Since weight is a force, it is measured in newtons, never kilograms. The relationship

$$W = mg \tag{1.5}$$

is used to calculate the weight in newtons given the mass in kilograms, and $g = 9.81$ m/s². Gravity is essentially constant on the earth's surface varying from 9.77 m/s² on the highest mountain to 9.83 m/s² in the deepest ocean trench.

Five of the nine basic dimensions and their units are included in Table 1.1; derived units of interest in our study of fluid mechanics are included in Table 1.2. Prefixes

Table 1.1   Basic Dimensions and Their Units

Quantity	Dimension	SI Units	English Units
Length l	L	meter m	foot ft
Mass m	M	kilogram kg	slug slug
Time t	T	second s	second sec
Temperature T	Θ	kelvin K	Rankine R
Plane angle		radian rad	radian rad

Table 1.2   Derived Dimensions and Their Units

Quantity	Dimension	SI Units	English Units
Area A	L^2	m^2	ft^2
Volume* $V\!\!\!\!-$	L^3	m^3 or L (liter)	ft^3
Velocity V	L/T	m/s	ft/sec
Acceleration a	L/T^2	m/s^2	ft/sec^2
Angular velocity Ω	T^{-1}	s^{-1}	sec^{-1}
Force F	ML/T^2	kg · m/s^2 or N	slug · ft/sec^2 or lb
Density ρ	M/L^3	kg/m^3	slug/ft^3
Specific weight γ	M/L^2T^2	N/m^3	lb/ft^3
Frequency f	T^{-1}	s^{-1}	sec^{-1}
Pressure p	M/LT^2	N/m^2 or Pa	lb/ft^2
Stress τ	M/LT^2	N/m^2 or Pa	lb/ft^2
Surface tension σ	M/T^2	N/m	lb/ft
Work W	ML^2/T^2	N · m or J	ft · lb
Energy E	ML^2/T^2	N · m or J	ft · lb
Heat rate $\dot{Q}$	ML^2/T^3	J/s	Btu/sec
Torque T	ML^2/T^2	N · m	ft · lb
Power $\dot{W}$	ML^2/T^3	J/s or W	ft · lb/sec
Mass flux $\dot{m}$	M/T	kg/s	slug/sec
Flow rate Q	L^3/T	m^3/s	ft^3/sec
Specific heat c	$L^2/T^2\Theta$	J/kg · K	Btu/slug · °R
Viscosity μ	M/LT	N · s/m^2	lb · sec/ft^2
Kinematic viscosity ν	L^2/T	m^2/s	ft^2/sec

*We use the special symbol $V\!\!\!\!-$ to denote volume and V to denote velocity.

Table 1.3   Prefixes for SI Units

Multiplication Factor	Prefix	Symbol
10^{12}	tera	T
10^{9}	giga	G
10^{6}	mega	M
10^{3}	kilo	k
10^{-2}	centi*	c
10^{-3}	mili	m
10^{-6}	micro	μ
10^{-9}	nano	n
10^{-12}	pico	p

*Discouraged except in cm, cm², or cm³.

are common in the SI system, so they are presented in Table 1.3. Note that the SI system is a special metric system. In our study we will use the units presented in these tables. We often use scientific notation, such as 3×10^5 N rather than 300 kN; either form is acceptable.

We finish this section with comments on significant figures. In almost every calculation, a material property is involved. Material properties are seldom known to four significant figures and often only to three. So, it is not appropriate to express answers to five or six significant figures. Our calculations are only as accurate as the least accurate number in our equations. For example, we use gravity as 9.81 m/s², only three significant figures. It is usually acceptable to express answers using four significant figures, but not five or six. The use of calculators may even provide eight. The engineer does not, in general, work with five or six significant figures. Note that if the leading digit in an answer is 1, it does not count as a significant figure, e.g., 12.48 has three significant figures.

EXAMPLE 1.1
Calculate the force needed to provide an initial upward acceleration of 40 m/s² to a 0.4-kg rocket.

Solution
Forces are summed in the vertical *y*-direction:

$$\sum F_y = ma_y$$
$$F - mg = ma$$
$$F - 0.4 \times 9.81 = 0.4 \times 40 \qquad \therefore F = 19.92 \text{ N}$$

Note that a calculator would provide 19.924 N, which contains four significant figures (the leading 1 doesn't count). Since gravity contained three significant figures, the 4 was dropped.

1.2 Gases and Liquids

The substance of interest in our study of fluid mechanics is a gas or a liquid. We restrict ourselves to those liquids that move under the action of a shear stress, no matter how small that shearing stress may be. All gases move under the action of a shearing stress but there are certain substances, like ketchup, that do not move until the shear becomes sufficiently large; such substances are included in the subject of rheology and are not presented in this book.

A force acting on an area is displayed in Fig. 1.1. A *stress vector* τ is the force vector divided by the area upon which it acts. The *normal stress* acts normal to the area and the *shear stress* acts tangent to the area. It is this shear stress that results in fluid motions. Our experience of a small force parallel to the water on a rather large boat confirms that any small shear causes motion. This shear stress is calculated with

$$\tau = \lim_{\Delta A \to 0} \frac{\Delta F_t}{\Delta A} \tag{1.6}$$

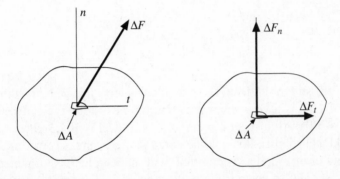

Figure 1.1   Normal and tangential components of a force.

Each fluid considered in our study is continuously distributed throughout a region of interest, that is, each fluid is a continuum. A liquid is obviously a continuum but each gas we consider is also assumed to be a continuum; the molecules are sufficiently close to one another so as to constitute a continuum. To determine if the molecules are sufficiently close, we use the *mean free path,* the average distance a molecule travels before it collides with a neighboring molecule. If the mean free path is small compared to a characteristic dimension of a device, the continuum assumption is reasonable. At high elevations, the continuum assumption is not reasonable and the theory of rarified gas dynamics is needed.

If a fluid is a continuum, the *density* can be defined as

$$\rho = \lim_{\Delta V \to 0} \frac{\Delta m}{\Delta V} \tag{1.7}$$

where Δm is the infinitesimal mass contained in the infinitesimal volume ΔV. Actually, the infinitesimal volume cannot be allowed to shrink to zero since near zero there would be few molecules in the small volume; a small volume ε would be needed as the limit in Eq. (1.7) for the definition to be acceptable. This is not a problem for most engineering applications, since there are 2.7×10^{16} molecules in a cubic millimeter of air at standard conditions. With the continuum assumption, quantities of interest are assumed to be defined at all points in a specified region. For example, the density is a continuous function of $x, y, z,$ and t, i.e., $\rho = \rho(x, y, z, t)$.

1.3 Pressure and Temperature

In our study of fluid mechanics, we often encounter pressure. It results from compressive forces acting on an area. In Fig. 1.2, the infinitesimal force ΔF_n acting on the infinitesimal area ΔA gives rise to the *pressure*, defined by

$$p = \lim_{\Delta A \to 0} \frac{\Delta F_n}{\Delta A} \tag{1.8}$$

The units on pressure result from force divided by area, that is, N/m², the pascal, Pa. A pressure of 1 Pa is a very small pressure, so pressure is typically expressed as kilopascals, or kPa. Atmospheric pressure at sea level is 101.3 kPa, or most often simply 100 kPa (14.7 psi). It should be noted that pressure is sometimes expressed as millimeters of mercury, as is common with meteorologists, or meters of water. We can use $p = \rho g h$ to convert the units, where ρ is the density of the fluid with height h.

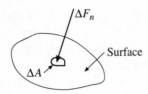

Figure 1.2   The normal force that results in pressure.

Pressure measured relative to atmospheric pressure is called *gage pressure*; it is what a gage measures if the gage reads zero before being used to measure the pressure. *Absolute pressure* is zero in a volume that is void of molecules, an ideal vacuum. Absolute pressure is related to gage pressure by the equation

$$p_{absolute} = p_{gage} + p_{atmosphere} \qquad (1.9)$$

where $p_{atmosphere}$ is the atmospheric pressure at the location where the pressure measurement is made. This atmospheric pressure varies considerably with elevation and is given in Table C.3. For example, at the top of Pikes Peak in Colorado, it is about 60 kPa. If neither the atmospheric pressure nor elevation are given, we will assume standard conditions and use $p_{atmosphere} = 100$ kPa. Figure 1.3 presents a graphic description of the relationship between absolute and gage pressure. Several common representations of the *standard atmosphere* (at 40° latitude at sea level) are included in that figure.

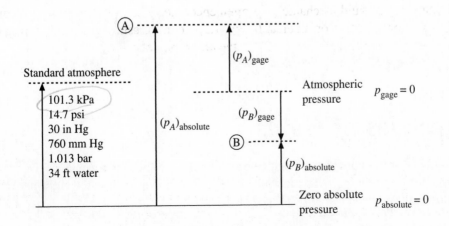

Figure 1.3   Absolute and gage pressure.

We often refer to a negative pressure, as at B in Fig. 1.3, as a vacuum; it is either a negative pressure or a *vacuum*. A pressure is always assumed to be a gage pressure unless otherwise stated. (In thermodynamics the pressure is assumed to be absolute.) A pressure of −30 kPa could be stated as 70 kPa absolute or a vacuum of 30 kPa, assuming atmospheric pressure to be 100 kPa (note that the difference between 101.3 kPa and 100 kPa is only 1.3 kPa, a 1.3% error, within engineering acceptability).

We do not define temperature (it requires molecular theory for a definition) but simply state that we use two scales: the Celsius scale and the Fahrenheit scale. The absolute scale when using temperature in degrees Celsius is the kelvin (K) scale. We use the conversion:

$$\underline{\quad\quad} \, K = \underline{\quad\quad} \, °C + 273.15 \tag{1.10}$$

In engineering problems we use the number 273, which allows for acceptable accuracy. Note that we do not use the degree symbol when expressing the temperature in degrees kelvin nor do we capitalize the word "kelvin." We read "100 K" as 100 kelvins in the SI system.

EXAMPLE 1.2

A pressure is measured to be a vacuum of 23 kPa at a location in Wyoming where the elevation is 3000 m. What is the absolute pressure?

Solution

P.239

Use Table C.3 to find the atmospheric pressure at 3000 m. We use a linear interpolation to find $p_{atmosphere} = 70.6$ kPa. Then

$$p_{abs} = p_{atm} + p = 70.6 - 23 = 47.6 \text{ kPa}$$

The vacuum of 23 kPa was expressed as −23 kPa in the equation.

1.4 Properties of Fluids

A number of fluid properties must be used in our study of fluid mechanics. Density, mass per unit volume, was introduced in Eq. (1.7). We often use weight per unit volume, the *specific weight* γ, related to density by

$$\gamma = \rho g \tag{1.11}$$

where g is the local gravity. For water γ is taken as 9810 N/m³ unless otherwise stated. Specific weight for gases is seldom used.

Specific gravity S is the ratio of the density of a substance to the density of water and is often specified for a liquid. It may be used to determine either the density or the specific weight:

$$\rho = S\rho_{\text{water}} \qquad \gamma = S\gamma_{\text{water}} \tag{1.12}$$

For example, the specific gravity of mercury is 13.6, which means that it is 13.6 times heavier than water. So, $\rho_{\text{mercury}} = 13.6 \times 1000 = 13\ 600$ kg/m³, where the density of water is 1000 kg/m³, the common value used for water.

Viscosity μ can be considered to be the internal stickiness of a fluid. It results in shear stresses in a flow and accounts for losses in a pipe or the drag on a rocket. It can be related in a one-dimensional flow to the velocity through a shear stress τ by

$$\tau = \mu\frac{du}{dr} \tag{1.13}$$

where we call *du/dr* a *velocity gradient*; r is measured normal to a surface and u is tangential to the surface, as in Fig. 1.4. Consider the units on the quantities in Eq. (1.13): the stress (force divided by an area) has units of N/m² so that the viscosity has the units N·s/m².

To measure the viscosity, consider a long cylinder rotating inside a second cylinder, as shown in Fig. 1.4. In order to rotate the inner cylinder with the rotational

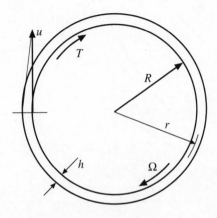

Figure 1.4   Fluid being sheared between two long cylinders.

speed Ω, a torque T must be applied. The velocity of the inner cylinder is $R\Omega$ and the velocity of the outer fixed cylinder is zero. The velocity distribution in the gap h between the cylinders is essentially a linear distribution as shown so that

$$\tau = \mu \frac{du}{dr} = \mu \frac{R\Omega}{h} \tag{1.14}$$

We can relate the shear to the applied torque as follows:

$$
\begin{aligned}
T &= \text{stress} \times \text{area} \times \text{moment arm} \\
&= \tau \times 2\pi RL \times R \\
&= \mu \frac{R\Omega}{h} \times 2\pi RL \times R = 2\pi \frac{R^3 \Omega L \mu}{h}
\end{aligned} \tag{1.15}
$$

where the shear stresses acting on the ends of the long cylinder have been neglected. A device used to measure the viscosity is a *viscometer*.

In an introductory course, attention is focused on *Newtonian fluids*, those that exhibit a linear relationship between the shear stress and the velocity gradient, as in Eqs. (1.13) and (1.14) and displayed in Fig. 1.5 (the normal coordinate here is y). Many common fluids, such as air, water, and oil are Newtonian fluids. *Non-Newtonian fluids* are classified as *dilatants, pseudoplastics,* and *ideal plastics* and are also displayed.

A very important effect of viscosity is to cause the fluid to stick to a surface, the *no-slip condition.* If a surface is moving extremely fast, as a satellite entering the atmosphere, this no-slip condition results in very large shear stresses on the surface; this results in extreme heat which can incinerate an entering satellite. The no-slip

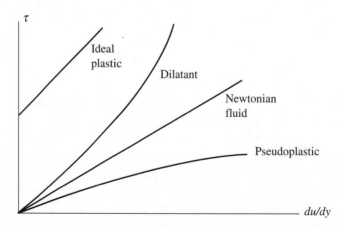

Figure 1.5 Newtonian and non-Newtonian fluids.

condition also gives rise to wall shear in pipes resulting in pressure drops that require pumps spaced appropriately over the length of a pipe line transporting a fluid such as oil or gas.

Viscosity is very dependent on temperature. Note that in Fig. C.1, the viscosity of a liquid decreases with increased temperature, but the viscosity of a gas increases with increased temperature. In a liquid, the viscosity is due to cohesive forces, but in a gas, it is due to collisions of molecules; both of these phenomena are insensitive to pressure. So we note that viscosity depends on only temperature in both a liquid and a gas, i.e., $\mu = \mu(T)$.

The viscosity is often divided by density in equations so we have defined the *kinematic viscosity* to be

$$v = \frac{\mu}{\rho} \tag{1.16}$$

It has units of m²/s. In a gas, we note that kinematic viscosity does depend on pressure since density depends on both temperature and pressure.

The volume of a gas is known to depend on pressure and temperature. In a liquid, the volume also depends slightly on pressure. If that small volume change (or density change) is important, we use the *bulk modulus B*, defined by

$$B = -V \left. \frac{\Delta p}{\Delta V} \right|_T = \rho \left. \frac{\Delta p}{\Delta \rho} \right|_T \tag{1.17}$$

The bulk modulus has the same units as pressure. It is included in Table C.1. For water at 20°C it is about 2100 MPa. To cause a 1% change in the volume of water, a pressure of 21 000 kPa is needed. So, it is obvious why we consider water to be incompressible. The bulk modulus is also used to determine the speed of sound c in water. It is given by

$$c = \sqrt{B/\rho} \tag{1.18}$$

This yields about $c = 1450$ m/s for water at 20°C.

Another property of occasional interest in our study is *surface tension σ*. It results from the attractive forces between molecules, and is included in Table C.1. It allows steel to float, droplets to form, and small droplets and bubbles to be spherical. Consider the free-body diagram of a spherical droplet and a bubble, as shown in Fig. 1.6. The pressure force inside the droplet balances the force due to surface tension around the circumference:

$$p\pi r^2 = 2\pi r\sigma \qquad \therefore p = \frac{2\sigma}{r} \tag{1.19}$$

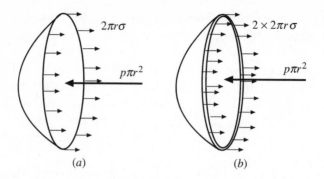

(a) (b)

Figure 1.6 Free-body diagrams of (a) a droplet and (b) a bubble.

Notice that in a bubble there are two surfaces so that the force balance provides

$$p = \frac{4\sigma}{r}$$ (1.20)

So, if the internal pressure is desired, it is important to know if it is a droplet or a bubble.

A second application where surface tension causes an interesting result is in the rise of a liquid in a capillary tube. The free-body diagram of the water in the tube is shown in Fig. 1.7. Summing forces on the column of liquid gives

$$\sigma \pi D \cos \beta = \rho g \frac{\pi D^2}{4} h$$ (1.21)

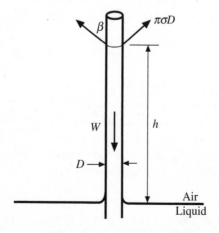

Figure 1.7 The rise of a liquid in a small tube.

where the right-hand side of the equation is the weight W. This provides the height the liquid will climb in the tube:

$$h = \frac{4\sigma \cos\beta}{\gamma D} \qquad (1.22)$$

The final property to be introduced in this section is vapor pressure. Molecules escape and reenter a liquid that is in contact with a gas, such as water in contact with air. The *vapor pressure* is that pressure at which there is equilibrium between the escaping and reentering molecules. If the pressure is below the vapor pressure, the molecules will escape the liquid; it is called *boiling* when water is heated to the temperature at which the vapor pressure equals the atmospheric pressure. If the local pressure is decreased to the vapor pressure, vaporization also occurs. This can happen when liquid flows through valves, elbows, or around turbine blades, should the pressure become sufficiently low; it is then called *cavitation*. The vapor pressure is found in Tables C.1 and C.5.

EXAMPLE 1.3
A 0.5 m × 2 m flat plate is towed at 5 m/s on a 2-mm-thick layer of SAE-30 oil at 38°C that separates it from a flat surface. The velocity distribution between the plate and the surface is assumed to be linear. What force is required if the plate and surface are horizontal?

Solution
The velocity gradient is calculated to be

$$\frac{du}{dy} = \frac{\Delta u}{\Delta y} = \frac{5-0}{0.002} = 2500 \text{ s}^{-1}$$

The force is the stress multiplied by the area:

$$F = \tau \times A = \mu \frac{du}{dy} \times A = 0.1 \times 2500 \times 0.5 \times 2 = 250 \text{ N}$$

Check the units to make sure the units of the force are newtons. The viscosity of the oil was found in Fig. C.1.

EXAMPLE 1.4
A machine creates small 1.0-mm-diameter bubbles of 20°C-water. Estimate the pressure that exists inside the bubbles.

Solution
Bubbles have two surfaces leading to the following estimate of the pressure:

$$p = \frac{4\sigma}{r} = \frac{4 \times 0.0736}{0.0005} = 589 \text{ Pa}$$

where the surface tension was taken from Table C.1.

1.5 Thermodynamic Properties and Relationships

A course in thermodynamics and/or physics usually precedes a fluid mechanics course. Those properties and relationships that are presented in those courses that are used in our study of fluids are included in this section. They are of particular use when compressible flows are studied, but they also find application to liquid flows.

We use the *ideal-gas law* in one of the two forms

$$p\mathcal{V} = mRT \qquad \text{or} \qquad p = \rho RT \qquad (1.23)$$

where the pressure p and the temperature T must be absolute quantities. The gas constant R is found in Table C.4.

Enthalpy is defined as

$$H = m\tilde{u} + p\mathcal{V} \qquad \text{or} \qquad h = \tilde{u} + p/\rho \qquad (1.24)$$

where $\tilde{u}$ is the *specific internal energy* (we use $\tilde{u}$ since u is used for a velocity component). In an ideal gas we can use

$$\Delta h = \int c_p dT \qquad \text{and} \qquad \Delta \tilde{u} = \int c_v dT \qquad (1.25)$$

where c_p and c_v are the specific heats also found in Table C.4. The specific heats are related to the gas constant by

$$c_p = c_v + R \qquad (1.26)$$

The ratio of specific heats is

$$k = \frac{c_p}{c_v} \qquad (1.27)$$

For liquids and solids, and for most gases over relatively small temperature differences, the specific heats are essentially constant and we can use

$$\Delta h = c_p \Delta T \qquad \text{and} \qquad \Delta \tilde{u} = c_v \Delta T \qquad (1.28)$$

For *adiabatic* (no heat transfer) *quasiequilibrium* (properties are constant throughout the volume at an instant) *processes* the following relationships can be used for an ideal gas assuming constant specific heats:

$$\frac{T_2}{T_1} = \left(\frac{p_2}{p_1}\right)^{(k-1)/k} = \left(\frac{\rho_2}{\rho_1}\right)^{k-1} \qquad \frac{p_2}{p_1} = \left(\frac{\rho_2}{\rho_1}\right)^{k} \qquad (1.29)$$

The adiabatic, quasiequilibrium process is also called an *isentropic process.*

A small pressure wave with a relatively low frequency travels through a gas with a wave speed of

$$c = \sqrt{kRT} \qquad (1.30)$$

Finally, the *first law of thermodynamics*, which we will refer to simply as *the energy equation*, will be of use in our study; it states that when a system, a fixed set of fluid particles, undergoes a change of state from state 1 to state 2, its energy changes from E_1 to E_2 as it exchanges energy with the surroundings in the form of work $W_{1\text{-}2}$ and heat transfer $Q_{1\text{-}2}$. This is expressed as

$$Q_{1\text{-}2} - W_{1\text{-}2} = E_2 - E_1 \qquad (1.31)$$

To calculate the heat transfer from given temperatures and areas, a course on heat transfer is required, so it is typically a given quantity in thermodynamics and fluid mechanics. The work, however, is a quantity that can often be calculated; it is a force times a distance and is often due to the pressure resulting in

$$W_{1\text{-}2} = \int_{l_1}^{l_2} F\,dl = \int_{l_1}^{l_2} pA\,dl = \int_{V_1}^{V_2} p\,dV \qquad (1.32)$$

The energy E considered in a fluids course consists of kinetic energy, potential energy, and internal energy:

$$E = m\left(\frac{V^2}{2} + gz + \tilde{u}\right) \qquad (1.33)$$

where the quantity in the parentheses is the specific energy e. If the properties are constant at an exit and an entrance to a flow, and there is no heat transferred and no losses, the above equation can be put in the form

$$\frac{V_2^2}{2g} + \frac{p_2}{\gamma_2} + z_2 = \frac{V_1^2}{2g} + \frac{p_1}{\gamma_1} + z_1 \qquad (1.34)$$

This equation does not follow directly from Eq. (1.31); it takes some effort to derive Eq. (1.34). An appropriate text could be consulted, but we will present it later in this book. It is presented here as part of our review of thermodynamics.

EXAMPLE 1.5
A farmer applies nitrogen to a crop from a tank pressurized to 1000 kPa absolute at a temperature of 25°C. What minimum temperature can be expected in the nitrogen if it is released to the atmosphere?

Solution
The minimum exiting temperature occurs for an isentropic process [see Eq. (1.29)]; it is

$$T_2 = T_1\left(\frac{p_2}{p_1}\right)^{(k-1)/k} = 298 \times \left(\frac{100}{1000}\right)^{0.4/1.4} = 154\ \text{K} \qquad \text{or} \qquad -119°\text{C}$$

Such a low temperature can cause serious injury should a hose break and the nitrogen impact the farmer.

Quiz No. 1

1. The correct units on viscosity are

    (A) $kg/(s \cdot m)$

    (B) $kg \cdot m/s$

    (C) $kg \cdot s/m$

    (D) $kg \cdot m/s^2$

2. The mean free path of a gas is $\lambda = 0.225 m/(\rho d^2)$ where d is the molecule's diameter, m is its mass, and ρ the density of the gas. Calculate the mean free path of air at sea level. For an air molecule $d = 3.7 \times 10^{-10}$ m and $m = 4.8 \times 10^{-26}$ kg.

    (A) 0.65×10^{-7} mm

    (B) 6.5×10^{-7} m

    (C) 65 nm

    (D) 650 µm

3. A vacuum of 25 kPa is measured at a location where the elevation is 4000 m. The absolute pressure, in millimeters of mercury, is nearest

    (A) 425 mm

    (B) 375 mm

    (C) 325 mm

    (D) 275 mm

4. The equation $p(z) = p_0 e^{-gz/RT_0}$ is a good approximation to the pressure in the atmosphere. Estimate the pressure at 6000 m using this equation and calculate the percent error assuming $p_0 = 100$ kPa and $T_0 = 15°C$. (The more accurate value is found in Table C.3.)

    (A) 4.4%

    (B) 4.0%

    (C) 3.2%

    (D) 2.6%

5. A fluid mass of 1500 kg occupies 2 m³. Its specific gravity is nearest

    (A) 1.5

    (B) 0.75

    (C) 0.30

    (D) 0.15

6. A viscometer is composed of two 12-cm-long, concentric cylinders with radii 4 and 3.8 cm. The outer cylinder is stationary and the inner one rotates. If a torque of 0.046 N·m is measured on the inner cylinder at a rotational speed of 120 rpm, estimate the viscosity of the liquid. Neglect the contribution to the torque from the cylinder ends and assume a linear velocity profile.

   (A) 0.127 N·s/m^2

   (B) 0.149 N·s/m^2

   (C) 0.161 N·s/m^2

   (D) 0.177 N·s/m^2

7. A 0.1-m^3 volume of water is observed to be 0.0982 m^3 after a pressure is applied. What is that pressure?

   (A) 37.8 MPa

   (B) 24.2 MPa

   (C) 11.7 MPa

   (D) 8.62 MPa

8. The pressure inside a 20-μm-diameter bubble of 20°C water is nearest

   (A) 17.8 kPa

   (B) 24.2 kPa

   (C) 29.4 kPa

   (D) 38.2 kPa

9. A car with tires pressurized to 270 kPa (40 psi) leaves Phoenix with the tire temperature at 60°C. Estimate the tire pressure (gage) when the car arrives in Alaska with a tire temperature of –30°C.

   (A) 270 kPa

   (B) 210 kPa

   (C) 190 kPa

   (D) 170 kPa

10. Air at 22°C is received from the atmosphere into a 200 cm^3 cylinder. Estimate the pressure and temperature (MPa, °C) if it is compressed isentropically to 10 cm^3.

   (A) (6.53, 705)

   (B) (5.72, 978)

   (C) (4.38, 978)

   (D) (7.43, 705)

11. Lightning is observed and thunder is heard 1.5 s later. About how far away did the lightning occur?

    (A) 620 m

    (B) 510 m

    (C) 430 m

    (D) 370 m

Quiz No. 2

1. If force, length, and time are selected as the three fundamental dimensions, what are the dimensions on mass?

2. A pressure of 28 kPa is measured at an elevation of 2000 m. What is the absolute pressure in mm of Hg?

3. Water at 20°C flows in a 0.8-cm-diameter pipe with a velocity distribution of $u(r) = 5[1 - (r^2/16) \times 10^6]$ m/s. Calculate the shear stress on the pipe wall.

4. SAE-30 oil at 30°C fills the gap between a 40-cm-diameter flat disk rotating 0.16 cm above a flat surface. Estimate the torque needed to rotate the disk at 600 rpm.

5. Water at 30°C is able to climb up a clean glass 0.2-mm-diameter tube due to surface tension. The water/glass angle is 0° with the vertical. How far up the tube does the water climb?

6. Derive an equation that relates the vertical force F needed to just lift a thin wire loop from a liquid assuming a vertical surface tension force. The wire radius is r and the loop diameter is D. Assume $D \gg r$.

7. A 2-m-long, 4 cm-diameter shaft rotates inside an equally long 4.02-cm-diameter cylinder. If SAE-10W oil at 25°C fills the gap between the concentric cylinders, estimate the horsepower needed to rotate the shaft at 1200 rpm assuming a linear velocity profile.

8. The *coefficient of thermal expansion* α_T allows the expansion of a liquid to be determined using the equation $\Delta V = \alpha_T V \Delta T$. What pressure is needed to cause the same decrease in volume of 2 m³ of 40°C water as that caused by a 10°C drop in temperature?

9. Calculate the weight of the column of air contained above a 1-m² area of atmospheric air from sea level to the top of the atmosphere.

10. Air expands from a tank maintained at 18°C and 250 kPa to the atmosphere. Estimate its minimum temperature as it exits.

CHAPTER 2

Fluid Statics

In *fluid statics* there is no relative motion between fluid particles so there are no shear stresses present (a shear results from a velocity gradient). This does not mean that the fluid particles are not moving, only that they are not moving relative to one another. If they are moving, as in a can of water rotating about its axis, they move as a solid body. The only stress involved in fluid statics is the normal stress, the pressure. It is the pressure acting over an area that gives rise to the force. Three types of problems are presented in this chapter: fluids at rest, as in the design of a dam; fluids undergoing linear acceleration, as in a rocket; and fluids that are rotating about an axis.

2.1 Pressure Variation

Pressure is a quantity that acts at a point. But, does it have the same magnitude in all directions at the point? To answer this question, consider Fig. 2.1. A pressure p is assumed to act on the hypotenuse and different pressures p_x and p_y on the other two sides of the infinitesimal element which has a uniform depth dz into the paper.

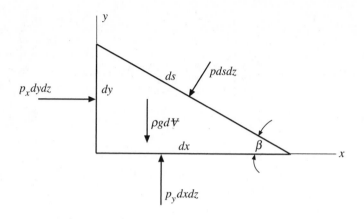

Figure 2.1 Pressure acting on an infinitesimal element.

The fluid particle occupying the fluid element could be accelerating so we apply Newton's second law in both the x- and y-directions:

$$\Sigma F_x = ma_x: \qquad p_x dy dz - p ds dz \sin\beta = \rho\frac{dx dy dz}{2}a_x$$

$$\Sigma F_y = ma_y: \qquad p_y dx dz - p ds dz \cos\beta - \rho g\frac{dx dy dz}{2} = \rho\frac{dx dy dz}{2}a_y$$

(2.1)

recognizing[1] that $d\math{V} = dx dy dz/2$. From Fig. 2.1 we have

$$dy = ds \sin\beta \qquad\qquad dx = ds \cos\beta \qquad\qquad (2.2)$$

Substitute these into Eq. (2.1) and obtain

$$p_x - p = \frac{\rho}{2}a_x dx$$

$$p_y - p = \frac{\rho}{2}(a_y + g)dy$$

(2.3)

Here we see that the quantities on the right-hand sides of the equations are infinitesimal (multiplied by dx and dy), and can be neglected[2] so that

$$p_x = p_y = p \qquad\qquad (2.4)$$

---

[1]We use the special symbol $\math{V}$ to represent volume and V to represent velocity.

[2]Mathematically, we could use an element with sides Δx and Δy and let $\Delta x \to 0$ and $\Delta y \to 0$.

Since the angle β is arbitrary, this holds for all angles. We could have selected dimensions dx and dz and arrived at $p_x = p_z = p$. So, in our applications to fluid statics, the pressure is a scalar function that acts equally in all directions at a point.

In the preceding discussion, pressure at a point was considered. But, how does pressure vary from point to point? The fluid element of depth dy in Fig. 2.2 is assumed to be accelerating. Newton's second law provides

$$pdydz - \left(p + \frac{\partial p}{\partial x} dx \right) dydz = \rho g dx dy dz a_x$$

$$pdxdy - \left(p + \frac{\partial p}{\partial z} dz \right) dxdy = -\rho g dx dy dz + \rho g dx dy dz a_z \tag{2.5}$$

If the element were shown in the y-direction also, the y-component equation would be

$$pdxdz - \left(p + \frac{\partial p}{\partial y} dy \right) dxdz = \rho g dx dy dz a_y \tag{2.6}$$

These three equations reduce to

$$\frac{\partial p}{\partial x} = -\rho a_x \qquad \frac{\partial p}{\partial y} = -\rho a_y \qquad \frac{\partial p}{\partial z} = -\rho(a_z + g) \tag{2.7}$$

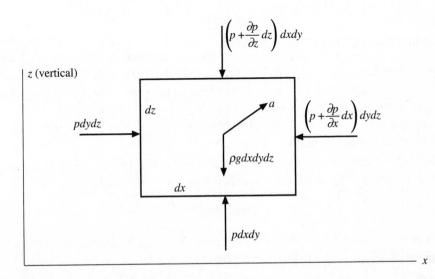

Figure 2.2 Forces acting on an element of fluid.

Finally, the pressure differential can be written as

$$dp = \frac{\partial p}{\partial x} dx + \frac{\partial p}{\partial y} dy + \frac{\partial p}{\partial z} dz$$
$$= -\rho a_x dx - \rho a_y dy - \rho(a_z + g)dz \qquad (2.8)$$

This can be integrated to give the desired difference in pressure between specified points in a fluid.

In a fluid at rest, there is no acceleration so the pressure variation from Eq. (2.8) is

$$dp = -\rho g dz \qquad \text{or} \qquad dp = -\gamma dz \qquad (2.9)$$

This implies that as the elevation z increases, the pressure decreases, a fact that we are aware of in nature; the pressure increases with depth in the ocean and decreases with height in the atmosphere.

If γ is constant, Eq. (2.9) allows us to write

$$\Delta p = -\gamma \Delta z \qquad (2.10)$$

where Δp is the pressure change over the elevation change Δz. If we desire an expression for the pressure, a distance h below a free surface (the pressure is zero on a free surface), it would be

$$p = \gamma h \qquad (2.11)$$

where $h = -\Delta z$. Equation (2.11) is used to convert pressure to an equivalent height of a liquid; atmospheric pressure is often expressed as millimeters of mercury (the pressure at the bottom of a 30-in column of mercury is the same as the pressure at the earth's surface due to the entire atmosphere).

If the pressure variation in the atmosphere is desired, Eq. (2.9) would be used with the ideal-gas law $p = \rho RT$ to give

$$dp = -\frac{p}{RT} g dz \qquad \text{or} \qquad \int_{p_0}^{p} \frac{dp}{p} = -\frac{g}{R} \int_{0}^{z} \frac{dz}{T} \qquad (2.12)$$

where p_0 is the pressure at $z = 0$. If the temperature could be assumed constant over the elevation change, this could be integrated to obtain

$$p = p_0 e^{-gz/RT} \qquad (2.13)$$

In the troposphere (between the earth's surface and an elevation of about 10 km), the temperature in kelvins is $T = 288 - 0.0065z$; Eq. (2.12) can be integrated to give the pressure variation.

EXAMPLE 2.1
Convert 230 kPa to millimeters of mercury, inches of mercury, and feet of water.

Solution
Equation (2.11) is applied using the specific weight of mercury as $13.6\gamma_{\text{water}}$:

$$p = \gamma h \qquad 230\,000 = (13.6 \times 9800) \times h$$

$$\therefore h = 1.726 \text{ m of mercury or } 1726 \text{ mm of mercury}$$

This is equivalent to $1.726 \text{ m} \times 3.281 \dfrac{\text{ft}}{\text{m}} \times 12 \dfrac{\text{in}}{\text{ft}} = 68.0$ in of mercury. Returning to Eq. (2.11), first convert kPa to psf:

$$230 \text{ kPa} \times 20.89 \frac{\text{psf}}{\text{kPa}} = 4805 \text{ psf} \qquad 4805 = 62.4 \times h$$

$$\therefore h = 77.0 \text{ ft of water}$$

We could have converted meters of mercury to feet of mercury and then multiplied by 13.6 to obtain feet of water.

2.2 Manometers

A *manometer* is an instrument that uses a column of a liquid to measure pressure. A typical U-tube manometer containing mercury is attached to a water pipe, as shown in Fig. 2.3. There are several ways to analyze a manometer. One way is to identify two points that have the same pressure, i.e., they are at the same elevation in the same liquid, such as points 2 and 3. Then we write

$$p_2 = p_3 \tag{2.14}$$
$$p_1 + \gamma_{\text{water}}h = p_4 + \gamma_{\text{Hg}}H$$

Since point 4 is shown to be open to the atmosphere, $p_4 = 0$. Thus, the manometer would measure the pressure p_1 in the pipe to be

$$p_1 = \gamma_{\text{Hg}}H - \gamma_{\text{water}}h \tag{2.15}$$

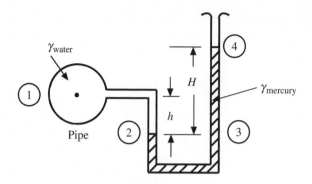

Figure 2.3   A U-tube manometer using water and mercury.

Some manometers will have several fluids with several interfaces. Each interface should be located with a point when analyzing a manometer.

EXAMPLE 2.2

A manometer connects an oil pipeline and a water pipeline, as shown. Determine the difference in pressure between the two pipelines using the readings on the manometer. Use $S_{oil} = 0.86$ and $S_{Hg} = 13.6$.

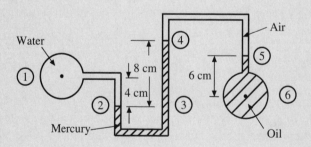

Solution

The points of interest have been positioned on the manometer shown. The pressure at point 2 is equal to the pressure at point 3:

$$p_2 = p_3$$
$$p_{water} + \gamma_{water} \times 0.04 = p_4 + \gamma_{Hg} \times 0.08$$

Observe that the heights must be in meters. The pressure at point 4 is essentially the same as that at point 5 since the specific weight of air is negligible compared to that of the oil. So,

$$p_4 = p_5$$
$$= p_{\text{oil}} - \gamma_{\text{oil}} \times 0.06$$

Finally,

$$p_{\text{water}} - p_{\text{oil}} = -\gamma_{\text{water}} \times 0.04 + \gamma_{\text{Hg}} \times 0.08 - \gamma_{\text{oil}} \times 0.06$$
$$= -9800 \times 0.04 + (13.6 \times 9800) \times 0.08 - (0.86 \times 9800) \times 0.06$$
$$= 9760 \text{ Pa}$$

2.3 Forces on Plane and Curved Surfaces

When a liquid is contained by a surface, such as a dam, the side of a ship, a water tank, or a levee, it is necessary to calculate the force and its location due to the liquid. The liquid is most often water but it could be oil or some other liquid. We will develop equations for forces on plane surfaces, but forces on curved surfaces can be determined using the same equations. Examples will illustrate.

Consider the general surface shown in Fig. 2.4. The liquid acts on the plane area shown as a section of the wall; a view from the top gives additional detail of the geometry. The force on the plane surface is due to the pressure $p = \gamma h$ acting over the area, that is,

$$F = \int_A p\,dA = \gamma \int_A h\,dA$$
$$= \gamma \sin \alpha \int_A y\,dA = \gamma \bar{y} A \sin \alpha \qquad (2.16)$$

where $\bar{y}$ is the distance[3] to the centroid of the plane area; the centroid is identified as the point C. Equation (2.16) can also be expressed as

$$F = \gamma \bar{h} A \qquad (2.17)$$

---

[3]Recall that $\bar{y} A = \int_A y\,dA$.

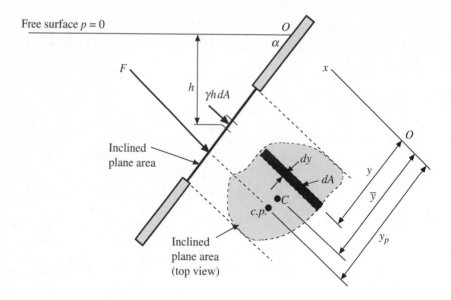

Figure 2.4   The force on an inclined plane area.

where $\bar{h}$ is the vertical distance to the centroid. Since $\gamma\bar{h}$ is the pressure at the centroid we see that the magnitude of the force is the area times the pressure that acts at the centroid of the area. It does not depend on the angle α of inclination. But, the force does not, in general, act at the centroid.

Let's assume that the force acts at some point called the *center of pressure (c.p.)*, located by the point (x_p, y_p). To determine where this point is, we must recognize that the sum of the moments of all the infinitesimal forces must equal the moment of the resultant force, that is,

$$y_p F = \gamma \int_A yh\, dA$$

$$= \gamma \sin\alpha \int_A y^2 dA = \gamma I_x \sin\alpha \qquad (2.18)$$

where I_x is the second moment[4] of the area about the x-axis. The parallel-axis-transfer theorem states that

$$I_x = \bar{I} + A\bar{y}^2 \qquad (2.19)$$

---

[4]Recall the second moment of a rectangle about its centroidal axis is $bh^3/12$.

where $\bar{I}$ is the moment of the area about its centroidal axis. So, substituting into Eq. (2.18) and using the expression for F from Eq. (2.16) results in

$$y_p = \bar{y} + \frac{\bar{I}}{A\bar{y}} \qquad (2.20)$$

This allows us to locate where the force acts. For a horizontal surface, the pressure is uniform over the area so that the pressure force acts at the centroid of the area. In general, y_p is greater than $\bar{y}$. The centroids and second moments of various areas are presented in books on statics or strength of materials. They will be given in the problems in this book.

If the top of the inclined area in Fig. 2.4 were at the free surface, the pressure distribution on that area would be triangular and the force F due to that pressure would act through the centroid of that triangular distribution, i.e., two-third the distance from the top of the inclined area.

To locate the x-coordinate x_p of the center of pressure, we use

$$x_p F = \gamma \sin\alpha \int_A xy\,dA$$
$$= \gamma I_{xy} \sin\alpha \qquad (2.21)$$

where I_{xy} is the product of inertia of the area. Using the transfer theorem for the product of inertia, the x-location of the center of pressure is

$$x_p = \bar{x} + \frac{\bar{I}_{xy}}{A\bar{y}} \qquad (2.22)$$

The above equations also allow us to calculate the forces acting on curved surfaces. Consider the curved gate shown in Fig. 2.5a. The objective in this problem would be to find the force P of the gate on the vertical wall and the forces F_x and F_y on the hinge. From the free-body diagrams in parts (b) and (c), the desired forces can be calculated providing the force F_W, which acts through the center of gravity of the area, can be found. The forces F_1 and F_2 can be found using Eq. (2.17). The forces F_H and F_V are the horizontal and vertical components of the force of the water acting on the gate. If a free-body diagram of only the water above the gate were identified, we would see that

$$F_H = F_1 \qquad \text{and} \qquad F_V = F_2 + F_W \qquad (2.23)$$

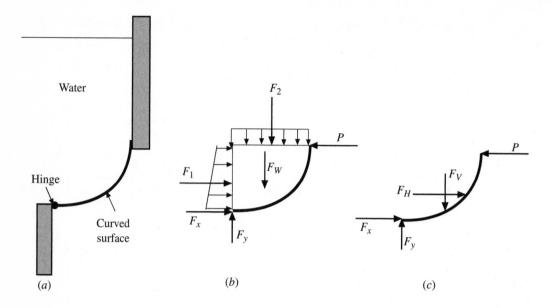

Figure 2.5 Forces on a curved surface: (*a*) the gate, (*b*) the water and the gate, and (*c*) the gate only.

Often, the gate is composed of a quarter circle. In this case, the problem can be greatly simplified by recognizing that the forces F_H and F_V, when added together as a vector, must act through the center of the quarter circle since all the infinitesimal forces due to the water pressure on the gate that make up F_H and F_V act through the center. So, for a gate that has the form of a part of a circle, the force components F_H and F_V can be located at the center of the circular arc. An example will illustrate.

A final application of forces on surfaces involves *buoyancy*, i.e., forces on floating bodies. *Archimedes' principle* states that there is a buoyancy force on a floating object equal to the weight of the displaced liquid, written as

$$F_B = \gamma V_{\text{displaced liquid}} \tag{2.24}$$

Since there are only two forces acting on a floating body, they must be equal and opposite and act through the center of gravity of the body (the body could have density variations) and the centroid of the liquid volume. The body would position itself so the center of gravity of the body and centroid of the liquid volume would be on a vertical line. Questions of stability arise (does the body tend to tip?) but are not considered here.

EXAMPLE 2.3

A 60-cm square gate has its top edge 12 m below the water surface. It is on a 45°
angle and its bottom edge is hinged as shown in (a). What force P is needed to
just open the gate?

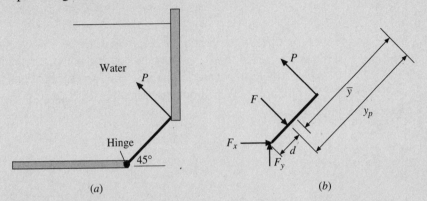

(a)   (b)

Solution

The first step is to sketch a free-body diagram of the gate so the forces and dis-
tances are clearly identified. It is provided in (b). The force F is calculated to be

$$F = \gamma \bar{h} A$$
$$= 9810 \times (12 + 0.3 \times \sin 45°) \times (0.6 \times 0.6) = 43\,130 \text{ N}$$

We will take moments about the hinge so it will not be necessary to calculate the
forces F_x and F_y. Let's find the distance d, the force F acts from the hinge:

$$\bar{y} = \frac{\bar{h}}{\sin 45°} = \frac{12 + 0.3 \sin 45°}{\sin 45°} = 17.27 \text{ m}$$

$$y_p = \bar{y} + \frac{\bar{I}}{A\bar{y}} = 17.27 + \frac{0.6 \times 0.6^3/12}{(0.6 \times 0.6) \times 17.27} = 17.272 \text{ m}$$

$$\therefore d = \bar{y} + 0.3 - y_p \cong 0.3 \text{ m}$$

Note: The distance $y_p - \bar{y}$ is very small and can be neglected because of the
relatively large 12 m height compared to the 0.6 m dimension. So, the force P is
calculated to be

$$P = \frac{0.3 \times F}{0.6} = 21\,600 \text{ N}$$

Note also that all dimensions are converted to meters.

EXAMPLE 2.4

Consider the gate shown to be a quarter circle of radius 80 cm with the hinge 8 m below the water surface (see Fig. 2.5). If the gate is 1 m wide, what force P is needed to hold the gate in the position shown?

Solution

Let's move the forces F_H and F_V to the center of the circular arc, as shown. This is allowed since all the infinitesimal force components that make up the resultant vector force $F_H + F_V$ pass through the center of the arc. The forces that act on the gate are displayed. If moments are taken about the hinge, F_x, F_y, and F_V produce no moments. So, there results

$$P = F_H$$

a rather simple result compared to the situation if we used Fig. 2.5c. The force P is

$$P = \gamma \bar{h} A = 9810 \times (8 - 0.4) \times (0.8 \times 1)$$
$$= 59\,600 \text{ N}$$

where $F_H = F_1$, and F_1 is the force on the vertical area shown in Fig. 2.5b.

2.4 Accelerating Containers

The pressure in a container accelerating with components a_x and a_y is found by integrating Eq. (2.8) between selected points 1 and 2 in Fig 2.6 to obtain

$$p_2 - p_1 = -\rho a_x (x_2 - x_1) - \rho(a_z + g)(z_2 - z_1) \qquad (2.25)$$

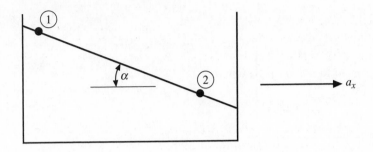

Figure 2.6 A linearly accelerating container.

If points 1 and 2 lie on a constant-pressure line (e.g., a free surface) such that $p_2 = p_1$, as in Fig. 2.6, and $a_z = 0$, Eq. (2.25) allows an expression for the angle α:

$$0 = -\rho a_x (x_2 - x_1) - \rho g (z_2 - z_1)$$

$$\tan \alpha = \frac{z_1 - z_2}{x_2 - x_1} = \frac{a_x}{g} \tag{2.26}$$

If a_z is not zero, it is simply included. The above equations allow us to make calculations involving linearly accelerating containers. The liquid is assumed to not be sloshing; it is moving as a rigid body. An example will illustrate.

To determine the pressure in a rotating container, Eq. (2.8) cannot be used so it is necessary to derive the expression for the differential pressure. Refer to the infinitesimal element of Fig. 2.7. A top view of the element is shown. Newton's second law applied in the radial r-direction provides, remembering that $a_r = r\Omega^2$,

$$prd\theta dz - (p + \frac{\partial p}{\partial r} dr)(r + dr)d\theta dz + pdrdz \sin\frac{d\theta}{2} + pdrdz \sin\frac{d\theta}{2} = \rho rd\theta drdz\, r\Omega^2 \tag{2.27}$$

Expand the second term carefully, use $\sin d\theta/2 = d\theta/2$, neglect higher order terms, and simplify Eq. (2.27) to

$$\frac{\partial p}{\partial r} = \rho r\Omega^2 \tag{2.28}$$

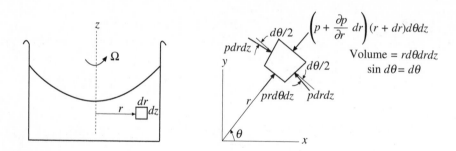

Figure 2.7   The rotating container and the top view of the infinitesimal element.

This provides the pressure variation in the radial direction and our usual $dp = -\rho g\,dz$ provides the pressure variation in the z-direction. Holding z fixed, the pressure difference from r_1 to r_2 is found by integrating Eq. (2.28):

$$p_2 - p_1 = \frac{\rho \Omega^2}{2}(r_2^2 - r_1^2) \qquad (2.29)$$

If point 1 is at the center of rotation so that $r_1 = 0$, then $p_2 = \rho \Omega^2 r_2^2/2$. If the distance from point 2 to the free surface (where $p_1 = 0$) is h as shown in Fig. 2.8, so that $p_2 = \rho g h$, we see that

$$h = \frac{\Omega^2 r_2^2}{2g} \qquad (2.30)$$

which is a parabola. The free surface is a paraboloid of revolution. An example illustrates the use of the above equations.

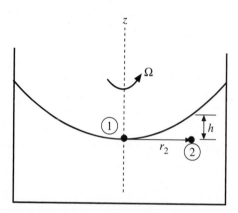

Figure 2.8   The free surface in a rotating container.

EXAMPLE 2.5
A 120-cm-long tank contains 80 cm of water and 20 cm of air maintained at 60 kPa above the water. The 60-cm-wide tank is accelerated at 10 m/s². After equilibrium is established, find the force acting on the bottom of the tank.

Solution
First, sketch the tank using the information given in the problem statement. It appears as shown.

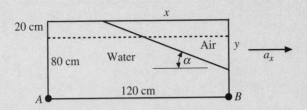

The distance x can be related to y by using Eq. (2.26):

$$\tan\alpha = \frac{a_x}{g} = \frac{10}{9.81} = \frac{y}{x} \qquad \therefore y = 1.019x$$

Equate the area of the air before and after to find x and y:

$$120 \times 20 = \frac{1}{2}xy = \frac{1.019}{2}x^2 \qquad \therefore x = 68.6 \text{ cm} \qquad \text{and} \qquad y = 69.9 \text{ cm}$$

The pressure will remain unchanged in the air above the water since the air volume doesn't change. The pressures at A and B are then [use Eq. (2.25)]:

$$p_A = 60\,000 + 1000 \times 10 \times (1.20 - 0.686) + 9810 \times 1.0 \text{ m} = 74\,900 \text{ Pa}$$
$$p_B = 60\,000 + 9810 \times (1.00 - 0.699) = 62\,900 \text{ Pa}$$

The average pressure on the bottom is $(p_A + p_B)/2$. Multiply the average pressure by the area to find the force acting on the bottom:

$$F = \frac{p_A + p_B}{2} \times A = \frac{74\,900 + 62\,900}{2} \times (1.2 \times 0.6) = 49\,600 \text{ N}$$

EXAMPLE 2.6

The cylinder shown is rotated about the center axis. What rotational speed is required so that the water just touches point A? Also, find the force on the bottom of the cylinder for that speed.

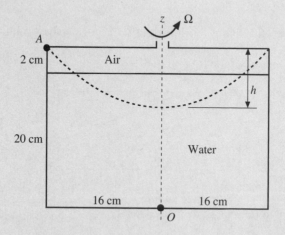

Solution

The volume of the air before and after must be the same. Recognizing that the volume of a paraboloid of revolution is half of the volume of a circular cylinder of the same radius and height, the height of the paraboloid of revolution is found:

$$\pi \times 0.16^2 \times 0.02 = \frac{1}{2}\pi \times 0.16^2 h \qquad \therefore h = 0.04 \text{ m}$$

Use Eq. (2.30) to find Ω:

$$0.04 = \frac{\Omega^2 \times 0.16^2}{2 \times 9.81} \qquad \therefore \Omega = 5.54 \text{ rad/s}$$

The pressure p on the bottom as a function of the radius r is given by

$$p - p_0 = \frac{\rho\Omega^2}{2}(r^2 - r_1^2) \text{ where } p_0 = 9810 \times (0.22 - 0.04) = 1766 \text{ Pa. So,}$$

$$p = \frac{1000 \times 5.54^2}{2} r^2 + 1766 = 15\,346r^2 + 1766$$

The pressure is integrated over the area to find the force to be

$$\int_0^{0.16}(15\,346r^2+1766)2\pi r\,dr = 2\pi\left(15\,346\times\frac{0.16^4}{4}+1766\times\frac{0.016^2}{2}\right)=157.8\text{ N}$$

Or, the force is simply the weight of the water:

$$F=\gamma V\!\!\!\!\!-\,=9810\times\pi\times0.16^2\times0.2=157.8\text{ N}$$

Quiz No. 1

1. Two meters of water is equivalent to how many millimeters of mercury?

   (A) 422 mm

   (B) 375 mm

   (C) 231 mm

   (D) 147 mm

2. A U-tube manometer measures the pressure in an air pipe to be 10 cm of water. The pressure in the pipe is nearest

   (A) 843 Pa

   (B) 981 Pa

   (C) 1270 Pa

   (D) 1430 Pa

3. Calculate the pressure in the water pipe if $h = 15$ cm and $H = 25$ cm.

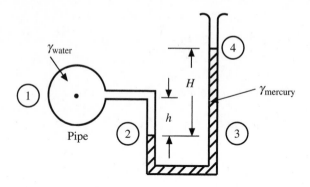

    (A) 22.8 kPa

    (B) 27.3 kPa

    (C) 31.9 kPa

    (D) 39.1 kPa

4. A submersible has a viewing window that is 60 cm in diameter. Determine the pressure force of the water on the window if the center of the window is 30 m below the surface and the horizontal window.

    (A) 65.8 kN

    (B) 79.3 kN

    (C) 83.2 kN

    (D) 99.1 kN

5. A 3-m-high, open cubical tank is filled with water. One end acts as a gate and has a hinge at the very bottom. What force at the very top of the gate is needed to just hold the gate shut?

    (A) 44.1 kN

    (B) 38.2 kN

    (C) 23.9 kN

    (D) 20.1 kN

6. The gate shown will open automatically when the water level reaches a
   certain height above the hinge. Determine that height if *b* is 1.6 m.

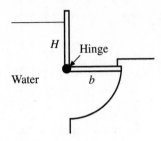

   (A) 1.87 m

   (B) 2.12 m

   (C) 2.77 m

   (D) 2.91 m

7. A body weighs 200 N in air and 125 N when submerged in water. Its
   specific weight is nearest

   (A) 2.31

   (B) 2.49

   (C) 2.54

   (D) 2.67

8. A 1.4-m-high, 4.2-m-long enclosed tank is filled with water and accelerated
   horizontally at 6 m/s². If the top of the tank has a small slit across the front,
   the maximum pressure in the tank is nearest

   (A) 38.9 kPa

   (B) 45.8 kPa

   (C) 59.7 kPa

   (D) 66.7 kPa

9. The force on the rear of the 80-cm-wide tank (the vertical end) of
   Prob. 8 is nearest

   (A) 108 kN

   (B) 95 kN

   (C) 79 kN

   (D) 56 kN

10. A test tube is placed in a rotating device that gradually positions the tube to a horizontal position. If the rate is 1000 rpm, estimate the pressure at the bottom of the relatively small-diameter test tube. The 12-cm-long tube contains water and the top of the tube is at a radius of 4 cm from the axis or rotation.

    (A) 723 kPa

    (B) 658 kPa

    (C) 697 kPa

    (D) 767 kPa

11. The U-tube is rotated about the right leg at 100 rpm. Calculate the pressures at A and B in the water if L is 40 cm.

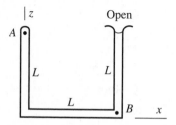

    (A) 6770 Pa, 3920 Pa

    (B) 7530 Pa, 2780 Pa

    (C) 8770 Pa, 3920 Pa

    (D) 9620 Pa, 2780 Pa

Quiz No. 2

1. The specific gravity of a liquid is 0.75. What height of that liquid is needed to provide a pressure difference of 200 kPa?

2. Assume a pressure of 100 kPa absolute at ground level. What is the pressure difference between the top of a 3-m-high wall on the outside where the temperature is −20°C and on the inside of a house where the temperature is 22°C? (This difference results in infiltration even if no wind is present.)

3. The pressure at the nose of a small airplane is given by $p = \rho V^2/2$, where ρ is the density of air. A U-tube manometer measures 25 cm of water. Determine the airplane's speed if it is flying at an altitude of 4000 m.

4. A submersible has a viewing window that is 60 cm in diameter. Determine the pressure force of the water on the window if the center of the window is 30 m below the surface and the window is at a 45° angle.

5. Find the force P needed to hold the 2-m-wide gate in the position shown if $h = 1.2$ m.

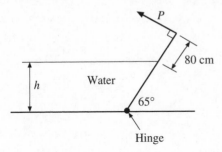

6. Find the force P needed to hold the 3-m-wide gate in the position shown if $r = 2$ m.

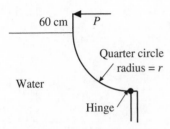

7. An object with a volume of 1200 cm³ weighs 20 N. What will it weigh when submerged in water?

8. A 1-m-high, 2-m-long enclosed tank is filled with water and accelerated horizontally at 6 m/s². If the top of the tank has a small slit across the back, calculate the distance from the front of the tank on the bottom where the pressure is zero.

9. Estimate the force acting on the bottom of the 80-cm-wide tank of Prob. 8.

10. Find the pressure at point A in the cylinder if $\omega = 100$ rpm and R is 60 cm.

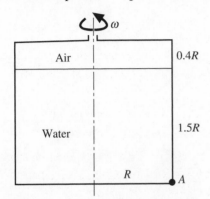

CHAPTER 3

Fluids in Motion

Fluid motions are quite complex and require rather advanced mathematics to describe them if all details are to be included. Simplifying assumptions can reduce the mathematics required, but even then the problems can get rather involved mathematically. To describe the motion of air around an airfoil, a tornado, or even water passing through a valve, the mathematics becomes quite sophisticated and is beyond the scope of an introductory course. We will, however, derive the equations needed to describe such motions, and make simplifying assumptions that will allow a number of problems of interest to be solved. These problems will include flow through a pipe and a channel, around rotating cylinders, and in a thin boundary layer near a flat wall. They will also include compressible flows involving simple geometries.

The pipes and channels will be straight and the walls perfectly flat. Fluids are all viscous, but often we can ignore the viscous effects. If viscous effects are to be included, we can demand that they behave in a linear fashion, a good assumption for water and air. Compressibility effects can also be ignored for low velocities such as those encountered in wind motions (including hurricanes) and flows around airfoils at speeds below about 100 m/s (220 mph).

In this chapter we will describe fluid motion in general, classify the different types of fluid motion, and also introduce the famous Bernoulli's equation along with its numerous assumptions.

3.1 Fluid Motion

LAGRANGIAN AND EULERIAN DESCRIPTIONS

Motion of a group of particles can be thought of in two basic ways: focus can be on an individual particle, such as following a particular car on a highway (a police patrol car may do this while moving with traffic), or it can be at a particular location as the cars move by (a patrol car sitting along the highway does this). When analyzed correctly, the solution to a problem would be the same using either approach (if you're speeding, you'll get a ticket from either patrol car).

When solving a problem involving a single object, such as in a dynamics course, focus is always on the particular object: the *Lagrangian description of motion*. It is quite difficult to use this description in a fluid flow where there are so many particles. Let's consider a second way to describe a fluid motion.

At a general point (x, y, z) in a flow, the fluid moves with a velocity $\mathbf{V}(x, y, z, t)$. The rate of change of the velocity of the fluid as it passes the point is $\partial\mathbf{V}/\partial t$, $\partial\mathbf{V}/\partial y$, $\partial\mathbf{V}/\partial z$, and it may also change with time at the point: $\partial\mathbf{V}/\partial t$. We use partial derivatives here since the velocity is a function of all four variables. This is the *Eulerian description of motion*, the description used in our study of fluids. We have used rectangular coordinates here but other coordinate systems, such as cylindrical coordinates, can also be used. The region of interest is referred to as a *flow field* and the velocity in that flow field is often referred to as the *velocity field*. The flow field could be the inside of a pipe, the region around a turbine blade, or the water in a washing machine.

If the quantities of interest using the Eulerian description were not dependent on time t, we would have a *steady flow*; the flow variables would depend only on the space coordinates. For such a flow,

$$\frac{\partial\mathbf{V}}{\partial t} = 0 \qquad \frac{\partial p}{\partial t} = 0 \qquad \frac{\partial\rho}{\partial t} = 0 \tag{3.1}$$

to list a few. In the above partial derivatives, it is assumed that the space coordinates remain fixed; we are observing the flow at a fixed point. If we followed a particular particle, as in a Lagrangian approach, the velocity of that particle would, in general, vary with time as it progressed through a flow field. But, using the Eulerian description, as in Eq. (3.1), time would not appear in the expressions for quantities in a steady flow, regardless of the geometry.

PATHLINES, STREAKLINES, AND STREAMLINES

Three different lines can be defined in a description of a fluid flow. The locus of points traversed by a particular fluid particle is a *pathline*; it provides the history of the particle. A time exposure of an illuminated particle would show a pathline. A *streakline* is the line formed by all particles passing a given point in the flow; it would be a snapshot of illuminated particles passing a given point. A *streamline* is a line in a flow to which all velocity vectors are tangent at a given instant; we cannot actually photograph a streamline. The fact that the velocity is tangent to a streamline allows us to write

$$\mathbf{V} \times d\mathbf{r} = 0 \tag{3.2}$$

since $\mathbf{V}$ and $d\mathbf{r}$ are in the same direction, as shown in Fig. 3.1; recall that two vectors in the same direction have a cross product of zero.

In a steady flow, all three lines are coincident. So, if the flow is steady, we can photograph a pathline or a streakline and refer to such a line as a streamline. It is the streamline in which we have primary interest in our study of fluids.

A *streamtube* is a tube whose walls are streamlines. A pipe is a streamtube, as is a channel. We often sketch a streamtube in the interior of a flow for derivation purposes.

ACCELERATION

To make calculations for a fluid flow, such as forces, it is necessary to describe the motion in detail; the expression for the acceleration is usually needed. Consider a fluid particle having a velocity $\mathbf{V}(t)$ at an instant t, as shown in Fig. 3.2. At the next instant, $t + \Delta t$, the particle will have velocity $\mathbf{V}(t + \Delta t)$, as shown. The acceleration of the particle is

$$\mathbf{a} = \frac{d\mathbf{V}}{dt} \tag{3.3}$$

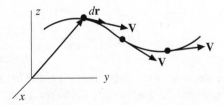

Figure 3.1   A streamline.

Fluid Mechanics Demystified

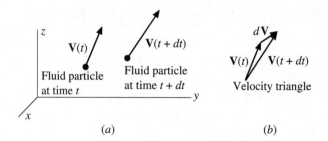

(a)   *(b)*

Figure 3.2   The velocity of a fluid particle.

where $d\mathbf{V}$ is shown in Fig 3.2(*b*). From the chain rule of calculus, we know that

$$d\mathbf{V} = \frac{\partial \mathbf{V}}{\partial x}dx + \frac{\partial \mathbf{V}}{\partial y}dy + \frac{\partial \mathbf{V}}{\partial z}dz + \frac{\partial \mathbf{V}}{\partial t}dt \tag{3.4}$$

since $\mathbf{V} = \mathbf{V}(x, y, z, t)$. This gives the acceleration as

$$\mathbf{a} = \frac{d\mathbf{V}}{dt} = \frac{\partial \mathbf{V}}{\partial x}\frac{dx}{dt} + \frac{\partial \mathbf{V}}{\partial y}\frac{dy}{dt} + \frac{\partial \mathbf{V}}{\partial z}\frac{dz}{dt} + \frac{\partial \mathbf{V}}{\partial t} \tag{3.5}$$

Now, since $\mathbf{V}$ is the velocity of a particle at (x, y, z), we let

$$\mathbf{V} = u\mathbf{i} + v\mathbf{j} + w\mathbf{k} \tag{3.6}$$

where (u, v, w) are the velocity components of the particle in the *x*-, *y*-, and *z*-directions, respectively, and $\mathbf{i}$, $\mathbf{j}$, and $\mathbf{k}$ are the unit vectors. For the particle at the point of interest, we have

$$\frac{dx}{dt} = u \qquad \frac{dy}{dt} = v \qquad \frac{dz}{dt} = w \tag{3.7}$$

so that the acceleration can be expressed as

$$\mathbf{a} = u\frac{\partial \mathbf{V}}{\partial x} + v\frac{\partial \mathbf{V}}{\partial y} + w\frac{\partial \mathbf{V}}{\partial z} + \frac{\partial \mathbf{V}}{\partial t} \tag{3.8}$$

In Eq. (3.8), the time derivative of velocity represents the *local acceleration* and the other three terms represent the *convective acceleration*. Local acceleration results if the velocity changes with time (e.g., startup), whereas convective acceleration results if velocity changes with position (as occurs at a bend or in a valve).

It is important to note that the expressions for the acceleration have assumed an inertial reference frame, i.e., the reference frame is not accelerating. It is assumed that a reference frame attached to the earth has negligible acceleration for problems of interest in this book. If a reference frame is attached to, say, a dishwasher spray arm, additional acceleration components, such as the Coriolis acceleration, enter the expression for the acceleration vector.

The vector equation (3.8) can be written as the three scalar equations:

$$a_x = u\frac{\partial u}{\partial x} + v\frac{\partial u}{\partial y} + w\frac{\partial u}{\partial z} + \frac{\partial u}{\partial t}$$

$$a_y = u\frac{\partial v}{\partial x} + v\frac{\partial v}{\partial y} + w\frac{\partial v}{\partial z} + \frac{\partial v}{\partial t} \tag{3.9}$$

$$a_z = u\frac{\partial w}{\partial x} + v\frac{\partial w}{\partial y} + w\frac{\partial w}{\partial z} + \frac{\partial w}{\partial t}$$

We write Eq. (3.3), and often Eq. (3.8), as

$$\mathbf{a} = \frac{D\mathbf{V}}{Dt} \tag{3.10}$$

where D/Dt is called the *material*, or *substantial derivative* since we have followed a material particle, or the substance. In rectangular coordinates, the material derivative is

$$\frac{D}{Dt} = u\frac{\partial}{\partial x} + v\frac{\partial}{\partial y} + w\frac{\partial}{\partial z} + \frac{\partial}{\partial t} \tag{3.11}$$

It can be used with other quantities of interest, such as the pressure: Dp/Dt would represent the rate of change of pressure of a fluid particle at some point (x, y, z).

The material derivative and acceleration components are presented for cylindrical and spherical coordinates in Table 3.1 at the end of this section.

ANGULAR VELOCITY AND VORTICITY

Visualize a fluid flow as the motion of a collection of fluid elements that deform and rotate as they travel along. At some instant in time, we could think of all the elements that make up the flow as being little cubes. If the cubes simply deform and don't rotate, we refer to the flow, or a region of the flow, as an *irrotational flow*.

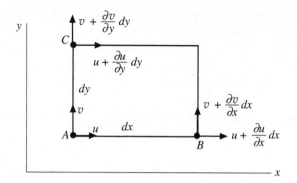

Figure 3.3 The rectangular face of a fluid element.

We are interested in such flows in our study of fluids; they exist in tornados away from the "eye" and in the flow away from the surfaces of airfoils and automobiles. If the cubes do rotate, they possess *vorticity*. Let's derive the equations that allow us to determine if a flow is irrotational or if it possesses vorticity.

Consider the rectangular face of an infinitesimal volume shown in Fig. 3.3. The *angular velocity* Ω_z about the z-axis is the average of the angular velocity of segments AB and AC, counterclockwise taken as positive:

$$\Omega_z = \frac{\Omega_{AB} + \Omega_{AC}}{2} = \frac{1}{2}\left(\frac{v_B - v_A}{dx} + \frac{-(u_C - u_A)}{dy}\right)$$

$$= \frac{1}{2}\left(\frac{\frac{\partial v}{\partial x}dx}{dx} - \frac{\frac{\partial u}{\partial y}dy}{dy}\right) = \frac{1}{2}\left(\frac{\partial v}{\partial x} - \frac{\partial u}{\partial y}\right) \tag{3.12}$$

If we select the other faces, we would find

$$\Omega_x = \frac{1}{2}\left(\frac{\partial w}{\partial y} - \frac{\partial v}{\partial z}\right) \qquad \Omega_y = \frac{1}{2}\left(\frac{\partial u}{\partial z} - \frac{\partial w}{\partial x}\right) \tag{3.13}$$

These three components of the angular velocity represent the rate at which a fluid particle rotates about each of the coordinate axes. The expression for Ω_z would predict the rate at which a cork would rotate in the *xy*-surface of the flow of water in a channel.

The *vorticity vector* ω is defined as twice the angular velocity vector: $\omega = 2\Omega$. The vorticity components are

$$\omega_x = \frac{\partial w}{\partial y} - \frac{\partial v}{\partial z} \qquad \omega_y = \frac{\partial u}{\partial z} - \frac{\partial w}{\partial x} \qquad \omega_z = \frac{\partial v}{\partial x} - \frac{\partial u}{\partial y} \tag{3.14}$$

Table 3.1   The Material Derivative and Acceleration in Cylindrical and Spherical Coordinates

Material Derivative	
Cylindrical	**Spherical**
$\dfrac{D}{Dt} = v_r\dfrac{\partial}{\partial r} + \dfrac{v_\theta}{r}\dfrac{\partial}{\partial \theta} + v_z\dfrac{\partial}{\partial z} + \dfrac{\partial}{\partial t}$	$\dfrac{D}{Dt} = v_r\dfrac{\partial}{\partial r} + \dfrac{v_\theta}{r}\dfrac{\partial}{\partial \theta} + \dfrac{v_\phi}{r\sin\theta}\dfrac{\partial}{\partial \phi} + \dfrac{\partial}{\partial t}$
Acceleration	
Cylindrical	**Spherical**
$a_r = v_r\dfrac{\partial v_r}{\partial r} + \dfrac{v_\theta}{r}\dfrac{\partial v_r}{\partial \theta} + v_z\dfrac{\partial v_r}{\partial z} - \dfrac{v_\theta^2}{r} + \dfrac{\partial v_r}{\partial t}$	$a_r = v_r\dfrac{\partial v_r}{\partial r} + \dfrac{v_\theta}{r}\dfrac{\partial v_r}{\partial \theta} + \dfrac{v_\phi}{r\sin\theta}\dfrac{\partial v_r}{\partial \phi} - \dfrac{v_\theta^2 + v_\phi^2}{r} + \dfrac{\partial v_r}{\partial t}$
$a_z = v_r\dfrac{\partial v_z}{\partial r} + \dfrac{v_\theta}{r}\dfrac{\partial v_z}{\partial \theta} + v_z\dfrac{\partial v_z}{\partial z} + \dfrac{\partial v_z}{\partial t}$	$a_\theta = v_r\dfrac{\partial v_\theta}{\partial r} + \dfrac{v_\phi}{r}\dfrac{\partial v_\theta}{\partial \theta} + \dfrac{v_\phi}{r\sin\theta}\dfrac{\partial v_\theta}{\partial \phi} + \dfrac{v_r v_\theta - v_\phi^2\cot\theta}{r} + \dfrac{\partial v_\theta}{\partial t}$
$a_\theta = v_r\dfrac{\partial v_\theta}{\partial r} + \dfrac{v_\theta}{r}\dfrac{\partial v_\theta}{\partial \theta} + v_z\dfrac{\partial v_\theta}{\partial z} + \dfrac{v_r v_\theta}{r} + \dfrac{\partial v_\theta}{\partial t}$	$a_\phi = v_r\dfrac{\partial v_\phi}{\partial r} + \dfrac{v_\theta}{r}\dfrac{\partial v_\phi}{\partial \theta} + \dfrac{v_\phi}{r\sin\theta}\dfrac{\partial v_\phi}{\partial \phi} + \dfrac{v_r v_\phi + v_\theta v_\phi\cot\theta}{r} + \dfrac{\partial v_\phi}{\partial t}$

EXAMPLE 3.1

A velocity field in a plane flow is given by $\mathbf{V} = 2yt\mathbf{i} + x\mathbf{j}$. Find the equation of the streamline passing through (4, 2) at $t = 2$.

Solution

Equation (3.2) can be written in the form

$$(2yt\mathbf{i} + x\mathbf{j}) \times (dx\mathbf{i} + dy\mathbf{j}) = (2yt\,dy - x\,dx)\mathbf{k} = \mathbf{0}$$

This leads to the equation, at $t = 2$,

$$4y\,dy = x\,dx$$

Integrate to obtain

$$4y^2 - x^2 = C$$

The constant is evaluated at the point (4, 2) to be $C = 0$. So, the equation of the streamline is

$$x^2 = 4y^2$$

Distance is usually measured in meters and time in seconds so that velocity would have units of m/s.

EXAMPLE 3.2

For the velocity field $\mathbf{V} = 2xy\mathbf{i} + 4tz^2\mathbf{j} - yz\mathbf{k}$, find the acceleration, the angular velocity about the z-axis, and the vorticity vector at the point $(2, -1, 1)$ at $t = 2$.

Solution

The acceleration is found, using $u = 2xy$, $v = 4tz^2$, and $w = -yz$, as follows:

$$\mathbf{a} = u\frac{\partial \mathbf{V}}{\partial x} + v\frac{\partial \mathbf{V}}{\partial y} + w\frac{\partial \mathbf{V}}{\partial z} + \frac{\partial \mathbf{V}}{\partial t}$$

$$= 2xy(2y\mathbf{i}) + 4tz^2(2x\mathbf{i} - z\mathbf{k}) - yz(8tz\mathbf{j} - y\mathbf{k}) + 4z^2\mathbf{j}$$

At the point $(2, -1, 1)$ and $t = 2$ there results

$$\mathbf{a} = 2(2)(-1)(-2\mathbf{i}) + 4(2)(1^2)(4\mathbf{i} - \mathbf{k}) - (-1)(1)(16\mathbf{j} + \mathbf{k}) + 4(1^2)\mathbf{j}$$

$$= 8\mathbf{i} + 32\mathbf{i} - 8\mathbf{k} + 16\mathbf{j} + \mathbf{k} + 4\mathbf{j}$$

$$= 40\mathbf{i} + 20\mathbf{j} - 7\mathbf{k}$$

The angular velocity component Ω_z is

$$\Omega_z = \frac{1}{2}\left(\frac{\partial v}{\partial x} - \frac{\partial u}{\partial y}\right) = \frac{1}{2}(0 - 2x) = x$$

At the point $(2, -1, 1)$ and $t = 2$, $\Omega_z = 2$. The vorticity vector is

$$\omega = \left(\frac{\partial w}{\partial y} - \frac{\partial v}{\partial z}\right)\mathbf{i} + \left(\frac{\partial u}{\partial z} - \frac{\partial w}{\partial x}\right)\mathbf{j} + \left(\frac{\partial v}{\partial x} - \frac{\partial u}{\partial y}\right)\mathbf{k}$$

$$= (-z - 8tz)\mathbf{i} + (0 - 0)\mathbf{j} + (0 - 2x)\mathbf{k}$$

At the point $(2, -1, 1)$ and $t = 2$, it is

$$\omega = (-1 - 16)\mathbf{i} - 4\mathbf{k} = -17\mathbf{i} - 4\mathbf{k}$$

Distance is usually measured in meters and time in seconds. Thus, angular velocity and vorticity would have units of m/s/m, i.e., rad/s.

3.2 Classification of Fluid Flows

Fluid mechanics is a subject in which many rather complicated phenomena are encountered, so it is important that we understand some of the descriptions and simplifications of several special fluid flows. Such special flows will be studied in detail in later chapters. Here we will attempt to classify them in as much detail as possible.

UNIFORM, ONE-, TWO-, AND THREE-DIMENSIONAL FLOWS

A dependent variable in our study of fluids depends, in general, on the three space coordinates and time, i.e., $V(x, y, z, t)$. The flow that depends on three space coordinates is a *three-dimensional flow*; it could be a steady flow if time is not involved, such as would be the case in the flow near the intersection of a wing and the fuselage of an aircraft flying at a constant speed. The flow in a washing machine would be an unsteady, three-dimensional flow.

Certain flows can be approximated as two-dimensional flows; flows over a wide weir, in the entrance region of a pipe, and around a sphere are examples that are of special interest. In such *two-dimensional flows* the dependent variables depend on only two space variables, i.e., $p(r, \theta)$ or $V(x, y, t)$. If the space coordinates are x and y, we refer to the flow as a *plane flow*.

One-dimensional flows are flows in which the velocity depends on only one space variable. They are of special interest in our introductory study since they include the flows in pipes and channels, the two most studied flows in an introductory course. For flow in a long pipe, the velocity depends on the radius r, and in a wide channel (parallel plates) it depends on y, as shown in Fig. 3.4.

The flows shown in Fig. 3.4 are also referred to as *developed flows*; the velocity profiles do not change with respect to the downstream coordinate. This demands that the pipe flow shown is many diameters downstream of any change in geometry, such as an entrance, a valve, an elbow, or a contraction or expansion. If the flow has not developed, the velocity field depends on more than one space coordinate, as is the case near a geometry change. The developed flow may be unsteady, i.e., it may depend on time, such as when a valve is being opened or closed.

Finally, there is the *uniform flow*, as shown in Fig. 3.5; the velocity profile, and other properties such as pressure, is uniform across the section of pipe. This profile

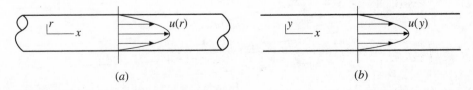

(a)   *(b)*

Figure 3.4   One-dimensional flow: (*a*) flow in a pipe and (*b*) flow in a wide channel.

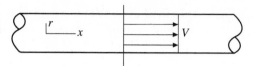

Figure 3.5   A uniform flow in a pipe.

is often assumed in pipe and channel flow problems since it approximates the more common turbulent flow so well. We will make this assumption in many of the problems throughout this book.

VISCOUS AND INVISCID FLOWS

In an *inviscid flow* the effects of viscosity can be completely neglected with no significant effects on the solution to a problem. All fluids have viscosity and if the viscous effects cannot be neglected, it is a *viscous flow*. Viscous effects are very important in pipe flows and many other kinds of flows inside conduits; they lead to losses and require pumps in long pipelines. But, there are flows in which we can neglect the influence of viscosity.

Consider an *external flow*, flow external to a body, such as the flow around an airfoil or a hydrofoil, as shown in Fig. 3.6. If the airfoil is moving relatively fast (faster than about 1 m/s), the flow away from a thin layer near the boundary, a *boundary layer*, can be assumed to have zero viscosity with no significant effect on the solution to the flow field (the velocity, pressure, temperature fields). All the viscous effects are concentrated inside the boundary layer and cause the velocity to be zero at the surface of the airfoil, the *no-slip condition*. Since inviscid flows are easier to solve than viscous flows, the recognition that the viscosity can be ignored in the flow away from the surface leads to much simpler solutions. This will be demonstrated in Chap. 8 (External Flows).

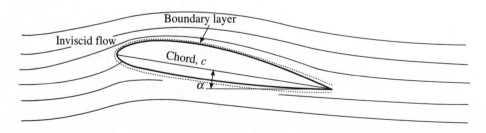

Figure 3.6   Flow around an airfoil.

LAMINAR AND TURBULENT FLOWS

A viscous flow is either a laminar flow or a turbulent flow. In a *turbulent flow* there is mixing of fluid particles so that the motion of a given particle is random and very irregular; statistical averages are used to specify the velocity, the pressure, and other quantities of interest. Such an average may be "steady" in that it is independent of time, or the average may be unsteady and depend on time. Figure 3.7 shows steady and unsteady turbulent flows. Notice the noisy turbulent flow from a faucet when you get a drink of water.

In a *laminar flow* there is negligible mixing of fluid particles; the motion is smooth and noiseless, like the slow flow of water from a faucet. If a dye is injected into a laminar flow, it remains distinct for a relatively long period of time. The dye would be immediately diffused if the flow were turbulent. Figure 3.8 shows a steady and an unsteady laminar flow. A laminar flow could be made to appear turbulent by randomly controlling a valve in the flow of honey in a pipe so as to make the velocity appear as in Fig. 3.7. Yet, it would be a laminar flow since there would be no mixing of fluid particles. So, a simple display of $V(t)$ is not sufficient to decide if a particular flow is laminar or turbulent. To be turbulent, the motion has to be random, as in Fig. 3.7, but it also has to have mixing of fluid particles.

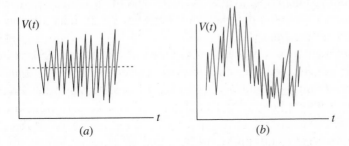

Figure 3.7   (*a*) Steady and (*b*) unsteady turbulent flows.

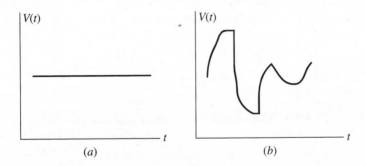

Figure 3.8   (*a*) Steady and (*b*) unsteady laminar flows.

As a flow starts from rest, as in a pipe when a valve is slightly opened, the flow starts out laminar, but as the average velocity increases, the laminar flow becomes unstable and turbulent flow ensues. In some cases, as in the flow between rotating cylinders, the unstable laminar flow develops into a secondary laminar flow of vortices, and then a third laminar flow, and finally a turbulent flow at higher speeds.

There is a quantity, called the *Reynolds number*, that is used to determine if a flow is laminar or turbulent. It is

$$\mathrm{Re} = \frac{VL}{\nu} \tag{3.15}$$

where V is a characteristic velocity (the average velocity in a pipe or the speed of an airfoil), L is a characteristic length (the diameter of a pipe or the distance from the leading edge of a flat plate), and ν is the kinematic viscosity. If the Reynolds number is larger than a critical Reynolds number, the flow is turbulent; if it is lower than a critical Reynolds number, the flow is laminar. For flow in a pipe, assuming the typically rough pipe wall, the critical Reynolds number is usually taken to be 2000; if the wall is smooth and free of vibrations, and the entering flow is free of disturbances, the critical Reynolds number can be as high as 40 000. The critical Reynolds number is different for each geometry. For flow between parallel plates, it is taken as 1500 using the average velocity and the distance between the plates. For a boundary layer on a flat plate, it is between 3×10^5 and 10^6, using L as the distance from the leading edge.

We do not refer to an inviscid flow as laminar or turbulent. In an external flow, the inviscid flow is called a *free-stream flow*. A free stream has disturbances but the disturbances are not accompanied by shear stresses, another requirement of both laminar and turbulent flows. The free stream can also be irrotational or it can possess vorticity.

A *boundary layer* is a thin layer of fluid that develops on a body due to the viscosity causing the fluid to stick to the boundary; it causes the velocity to be zero at the wall. The viscous effects in such a layer can actually incinerate a satellite on reentry. Figure 3.9 shows the typical boundary layer on a flat plate. It is laminar near the leading edge and undergoes transition to a turbulent flow with sufficient

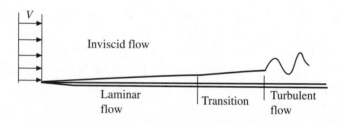

Figure 3.9 Boundary-layer flow on a flat plate.

length. For a smooth rigid plate with low free-stream fluctuation level, a laminar layer can exist up to Re = 10^6, L being the length along the plate; for a rough plate, or a vibrating plate, or high free-stream fluctuations, a laminar flow exists up to about Re = 3×10^5.

INCOMPRESSIBLE AND COMPRESSIBLE FLOWS

Liquid flows are assumed to be incompressible in most situations (water hammer[1] is an exception). In such *incompressible flows* the density of a fluid particle as it moves along is assumed to be constant, i.e.,

$$\frac{D\rho}{Dt} = 0 \tag{3.16}$$

This does not demand that the density of all the fluid particles be the same. For example, salt could be added to a water flow at some point in a pipe so that downstream of the point the density would be greater than at some upstream point. Atmospheric air at low speeds is incompressible but the density decreases with increased elevation, i.e., $\rho = \rho(z)$, where z is vertical. We usually assume a fluid to have constant density when we make the assumption of incompressibility. Constant density requires

$$\frac{\partial \rho}{\partial t} = 0 \qquad \frac{\partial \rho}{\partial x} = 0 \qquad \frac{\partial \rho}{\partial y} = 0 \qquad \frac{\partial \rho}{\partial z} = 0 \tag{3.17}$$

The flow of air can be assumed to be incompressible if the velocity is sufficiently low. Air flow in conduits, around automobiles and small aircraft, and the takeoff and landing of commercial aircraft are all examples of incompressible airflows. The *Mach number* M where

$$M = \frac{V}{c} \tag{3.18}$$

is used to determine if a flow is compressible; V is the characteristic velocity and $c = \sqrt{kRT}$ is the speed of sound. If M < 0.3, we assume the flow to be incompressible. For air near sea level this is about 100 m/s (200 mph) so many air flows can be assumed to be incompressible. Compressibility effects are considered in some detail in Chap. 9 (Compressible Flows).

---

[1]Water hammer may occur when a sudden change occurs in a flow, such as a sudden closing of a valve.

EXAMPLE 3.3
A river flowing through campus appears quite placid. A leaf floats by and we estimate that it floats about 2 m in 10 s. We wade in the water and estimate the depth to be about 60 cm. Is the flow laminar or turbulent?

Solution
We estimate the Reynolds number to be, assuming $T = 20°C$ and using Table C.1,

$$\text{Re} = \frac{Vh}{\nu} = \frac{(2/10) \times 0.6}{10^{-6}} = 120\,000$$

The flow is highly turbulent at this Reynolds number, contrary to our observation of the placid flow. Most internal flows are turbulent, as observed when we drink from a drinking fountain. Laminar flows are of minimal importance to engineers when compared to turbulent flows; a lubrication problem is one exception.

3.3 Bernoulli's Equation

Bernoulli's equation may be the most often used equation from fluid mechanics, but it is also the most often misused equation. In this section it will be derived and the restrictions required for its derivation will be highlighted. But, before the equation is derived, consider the five assumptions required: negligible viscous effects (no shear stresses), constant density, steady flow, the flow is along a streamline, and the velocity is measured in an inertial reference frame.

We apply Newton's second law to a cylindrical particle that is moving on a streamline, as shown in Fig. 3.10. A summation of infinitesimal forces acting on the particle is

$$pdA - \left(p + \frac{\partial p}{\partial s} ds \right) dA - \rho g ds dA \cos\theta = \rho ds dA a_s \qquad (3.19)$$

where a_s is the s-component of the acceleration vector. It is given by Eq. (3.9) where we think of the x-direction being in the s-direction so that $u = V$:

$$a_s = V \frac{\partial V}{\partial s} + \frac{\partial V}{\partial t} \qquad (3.20)$$

where $\partial V/\partial t = 0$, assuming a steady flow. (This leads to the same acceleration expression as presented in physics or dynamics where $a_x = VdV/dx$, providing an

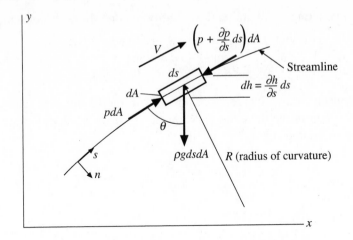

Figure 3.10   A particle moving along a streamline.

inertial reference frame is used in which no Coriolis or other acceleration components are present.) Next, we observe that

$$dh = ds\cos\theta = \frac{\partial h}{\partial s}ds \qquad (3.21)$$

resulting in

$$\cos\theta = \frac{\partial h}{\partial s} \qquad (3.22)$$

Now, divide Eq. (3.19) by $ds\,dA$ and use the above expressions for a_s and $\cos\theta$ and rearrange. There results

$$-\frac{\partial p}{\partial s} - \rho g\frac{\partial h}{\partial s} = \rho V\frac{\partial V}{\partial s} \qquad (3.23)$$

If we assume that the density ρ is constant (this is more restrictive than incompressibility) so it can be moved after the partial derivative, and we recognize that $V\partial V/\partial s = \partial(V^2/2)/\partial s$, we can write our equation as

$$\frac{\partial}{\partial s}\left(\frac{V^2}{2g} + \frac{p}{\rho g} + h\right) = 0 \qquad (3.24)$$

This means that along a streamline the quantity in parentheses is constant, i.e.,

$$\frac{V^2}{2g} + \frac{p}{\rho g} + h = \text{const} \tag{3.25}$$

where the constant may change from one streamline to the next; along a given streamline the sum of the three terms is constant. This is often written, referring to two points on the same streamline, as

$$\frac{V_1^2}{2g} + \frac{p_1}{\rho g} + h_1 = \frac{V_2^2}{2g} + \frac{p_2}{\rho g} + h_2 \tag{3.26}$$

or

$$\frac{V_1^2}{2} + \frac{p_1}{\rho} + gh_1 = \frac{V_2^2}{2} + \frac{p_2}{\rho} + gh_2 \tag{3.27}$$

Either of the two forms above is the famous *Bernoulli's equation*. Let's once again state the assumptions required to use Bernoulli's equation:

- Inviscid flow (no shear stresses)
- Constant density
- Steady flow
- Along a streamline
- Applied in an inertial reference frame

The first three listed are primarily ones usually considered. There are special applications where the last two must be taken into account; but those special applications will not be presented in this book. Also, we often refer to a constant-density flow as an incompressible flow, even though constant density is more restrictive [refer to the comments after Eq. (3.16)]. This is because incompressible flows, in which the density changes from one streamline to the next, such as in atmospheric flows, are not encountered in an introductory course.

Note that the units on all the terms in Eq. (3.26) are meters. Consequently, $V^2/2g$ is called the *velocity head*, $p/\rho g$ is the *pressure head*, and h is simply the *head*. The sum of the three terms is often referred to as the *total head*. The pressure p is the *static pressure* and the sum $p + \rho V^2/2$ is the *total pressure* or *stagnation pressure* since it is the pressure at a *stagnation point*, a point where the fluid is brought to rest along a streamline.

The difference in the pressures can be observed by considering the measuring probes shown in Fig. 3.11. The probe in Fig. 3.11*a* is a *piezometer*; it measures the

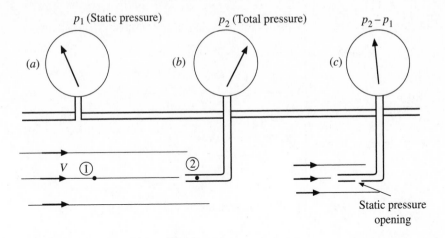

Figure 3.11   Pressure probes: (*a*) the piezometer, (*b*) a pitot tube, and (*c*) a pitot-static tube.

static pressure, or simply, the pressure at point 1. The *pitot tube* in Fig. 3.11*b* measures the total pressure, the pressure at a point where the velocity is zero, as at point 2. And, the *pitot-static tube*, which has a small opening the side of the probe as shown in Fig. 3.11*c*, is used to measure the difference between the total pressure and the static pressure, i.e., $\rho V^2/2$; this is used to calculate the velocity. The expression for velocity is

$$V = \sqrt{\frac{2}{\rho}(p_2 - p_1)} \qquad (3.28)$$

where point 2 must be a stagnation point with $V_2 = 0$. So, if only the velocity is desired, we simply use the pitot-static probe shown in Fig. 3.11*c*.

Bernoulli's equation is used in numerous fluid flows. It can be used in an internal flow over short reaches if the viscous effects can be neglected; such is the case in the well-rounded entrance to a pipe (see Fig. 3.12) or in a rather sudden contraction of a pipe. The velocity for such an entrance is approximated by Bernoulli's equation to be

$$V_2 = \sqrt{\frac{2}{\rho}(p_1 - p_2)} \qquad (3.29)$$

Another common application of the Bernoulli's equation is from the free stream to the front area of a round object such as a sphere or a cylinder or an airfoil. A sketch

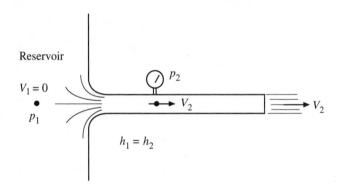

Figure 3.12 Flow from a reservoir through a pipe.

is helpful, as shown in Fig. 3.13. For many flow situations the flow separates from the surface, resulting in a separated flow, as shown. If the flow approaching the object is uniform, the constant in Eq. (3.25) will be the same for all the streamlines. Bernoulli's equation can then be applied from the free stream to the stagnation point at the front of the object, and to points along the surface of the object up to the separation region.

We often solve problems involving a pipe exiting to the atmosphere. For such a situation the pressure just inside the pipe exit is the same as the atmospheric pressure just outside the pipe exit, since the streamlines exiting the pipe are straight near the exit (see Fig. 3.12). This is quite different from the entrance flow of Fig. 3.12 where the streamlines near the entrance are extremely curved.

To approximate the pressure variation normal to a curved streamline, consider the particle of Fig. 3.10 to be a parallelepiped with thickness dn normal to the streamline, with area dA_s of the side, and with length ds. Use $\Sigma F_n = ma_n$:

$$pdA_s - \left(p + \frac{\partial p}{\partial n}dn\right)dA_s - \rho g\, dn\, dA_s = \rho\, dn\, dA_s \frac{V^2}{R} \qquad (3.30)$$

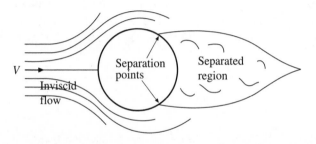

Figure 3.13 Flow around a sphere or a long cylinder.

where we have used the acceleration to be V^2/R, R being the radius of curvature in the assumed plane flow. If we assume that the effect of gravity is small compared to the acceleration term, this equation simplifies to

$$-\frac{\partial p}{\partial n} = \rho \frac{V^2}{R} \qquad (3.31)$$

Since we will use this equation to make estimations of pressure changes normal to a streamline, we approximate $\partial p/\partial n = \Delta p/\Delta n$ and arrive at the relationship

$$-\frac{\Delta p}{\Delta n} = \rho \frac{V^2}{R} \qquad (3.32)$$

Hence, we see that the pressure decreases as we move toward the center of the curved streamlines; this is experienced in a tornado where the pressure can be extremely low in the tornado's "eye." This reduced pressure is also used to measure the intensity of a hurricane; that is, lower the pressure in the hurricane's center, larger the velocity at its outer edges.

EXAMPLE 3.4
The wind in a hurricane reaches 200 km/h. Estimate the force of the wind on a window facing the wind in a high-rise building if the window measures 1 m by 2 m. Use the density of the air to be 1.2 kg/m³.

Solution
Use Bernoulli's equation to estimate the pressure on the window:

$$p = \rho \frac{V^2}{2} = 1.2 \times \frac{(200 \times 1000/3600)^2}{2} = 1852 \text{ N/m}^2$$

where the velocity must have units of m/s. To check the units, use kg = N·s²/m. Assume the pressure to be essentially constant over the window so that the force is then

$$F = pA = 1852 \times 1 \times 2 = 3704 \text{ N} \qquad \text{or} \qquad 833 \text{ lb}$$

This force is large enough to break many windows, especially if they are not properly designed.

EXAMPLE 3.5

A piezometer is used to measure the pressure in a pipe to be 20 cm of water. A pitot tube measures the total pressure to be 33 cm of water at the same general location. Estimate the velocity of the water in the pipe.

Solution

The velocity is found using Eq. (3.27):

$$V = \sqrt{\frac{2}{\rho}(p_2 - p_1)} = \sqrt{2g(h_2 - h_1)} = \sqrt{2 \times 9.81 \times (0.33 - 0.20)} = 1.60 \text{ m/s}$$

where we used the pressure relationship $p = \rho g h$.

Quiz No. 1

1. A velocity field in a plane flow is given by $\mathbf{V} = 2yt\mathbf{i} + x\mathbf{j}$ m/s. The magnitude of the acceleration at the point (4, 2 m) at $t = 3$ s is

    (A) 52.5 m/s^2

    (B) 48.5 m/s^2

    (C) 30.5 m/s^2

    (D) 24.5 m/s^2

2. A velocity field in a plane flow is given by $\mathbf{V} = 2xy\mathbf{i} + yt\mathbf{j}$ m/s. The vorticity of the fluid at the point (0, 4 m) at $t = 3$ s is

    (A) $-4\mathbf{k}$ rad/s

    (B) $-3\mathbf{j}$ rad/s

    (C) $-2\mathbf{k}$ rad/s

    (D) $-3\mathbf{i}$ rad/s

3. The parabolic velocity distribution in a channel flow is given by $u(y) = 0.2(1 - y^2)$ m/s, with y measured in centimeters. What is the acceleration of a fluid particle at a location where $y = 0.5$ cm?

    (A) 0

    (B) 2 m/s^2

    (C) 4 m/s^2

    (D) 5 m/s^2

4. The equation of the streamline that passes through the point $(2, -1)$ if the velocity field is given by $\mathbf{V} = 2xy\mathbf{i} + y^2\mathbf{j}$ m/s is

   (A) $x^2 = 4y^2$

   (B) $x = -2y$

   (C) $x^2 = -4y$

   (D) $x = 2y^2$

5. A drinking fountain has a 2-mm-diameter outlet. If the water is to be laminar, what is the maximum speed that the water should have?

   (A) 0.5 m/s

   (B) 1 m/s

   (C) 2 m/s

   (D) 4 m/s

6. Which of the following flows could be modeled as inviscid flows?

   (a) Developed flow in a pipe

   (b) Flow of water over a long weir

   (c) Flow in a long, straight canal

   (d) The flow of exhaust gases exiting a rocket

   (e) Flow of blood in an artery

   (f) Flow of air around a bullet

   (g) Flow of air in a tornado

   (A) d, e

   (B) d, g

   (C) b, e

   (D) b, g

7. Salt is being added to fresh water in a pipe at a certain location. In the vicinity of that location the term $D\rho/Dt$ is nonzero. Which term in the expression for $D\rho/Dt$ is nonzero if the x-axis is along the pipe axis? Assume uniform conditions.

   (A) $u\partial\rho/\partial y$

   (B) $\rho\partial u/\partial x$

   (C) $u\partial\rho/\partial x$

   (D) $\rho\partial u/\partial y$

8. A pitot probe measures 10 cm of water on a small airplane flying where the temperature is 20°C. The speed of the airplane is nearest

   (A) 40 m/s

   (B) 50 m/s

   (C) 60 m/s

   (D) 70 m/s

9. The fluid in the pipe is water and $h = 10$ cm of mercury. The velocity V is nearest

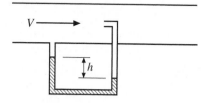

   (A) 7 m/s

   (B) 8 m/s

   (C) 5 m/s

   (D) 4 m/s

10. Select the false statement for Bernoulli's equation

    (A) It can be applied to an inertial coordinate system.

    (B) It can be applied to an unsteady flow.

    (C) It can be applied in an inviscid flow.

    (D) It can be applied between two points along a streamline.

11. Water flows through a long-sweep elbow on a 2-cm-diameter pipe at an average velocity of 20 m/s. Estimate the increase in pressure from the inside of the pipe to the outside of the pipe midway through the elbow if the radius of curvature of the elbow averages 12 cm at the midway section.

    (A) 60 kPa

    (B) 66.7 kPa

    (C) 75 kPa

    (D) 90 kPa

Quiz No. 2

1. Find the rate of change of the density in a stratified flow where
   $\rho = 1000(1 - 0.2z)$ and the velocity is $\mathbf{V} = 10(z - z^2)\mathbf{i}$.

2. A velocity field is given in cylindrical coordinates as

$$v_r = \left(2 - \frac{8}{r^2}\right)\cos\theta \text{ m/s}, \quad v_\theta = -\left(2 + \frac{8}{r^2}\right)\sin\theta \text{ m/s}, \quad v_z = 0$$

   What are the three acceleration components at the point (3 m, 90°)?

3. The traffic in a large city is to be studied. Explain how it would be done
   using (a) the Lagrangian approach and (b) the Eulerian approach.

4. Find the unit vector normal to the streamline at the point (2, 1) when $t = 2$ s
   if the velocity field is given by $\mathbf{V} = 2xy\mathbf{i} + y^2 t\mathbf{j}$ m/s.

5. A leaf is floating in a river seemingly quite slowly. It is timed to move 6 m
   in 40 s. If the river is about 1.2 m deep, determine if the placidly flowing
   river is laminar or turbulent.

6. Which of the following flows would definitely be modeled as a turbulent flow?

   (a) Developed flow in a pipe

   (b) Flow of water over a long weir

   (c) Flow in a long, straight canal

   (d) The flow of exhaust gases exiting a rocket

   (e) Flow of blood in an artery

   (f) Flow of air around a bullet

   (g) Flow of air in a tornado

7. Air flows over and parallel to a 10-m-long flat plate at 2 m/s. How long is
   the laminar portion of the boundary layer if the air temperature is 30°C.
   Assume a high-fluctuation level on a smooth rigid plate.

8. The pitot and piezometer probes read the total and static pressures as shown. Calculate the velocity V.

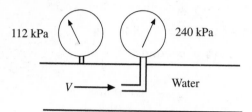

9. Determine the velocity V in the pipe if water is flowing and $h = 20$ cm of mercury.

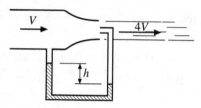

10. A car is travelling at 120 km/h. Approximate the force of the air on the 20-cm-diameter flat lens on the headlight.

CHAPTER 4

The Integral Equations

Many, if not most, of the quantities of interest in fluid mechanics are integral quantities; they are found by integrating some property of interest over an area or a volume. Many times the property is essentially constant so the integration is easily performed but other times, the property varies over the area or volume, and the required integration may be quite difficult. Some of the integral quantities of interest are: the rate of flow through a pipe, the kinetic energy in the wind approaching a wind machine, the power generated by the blade of a turbine, and the drag on an airfoil. There are quantities that are not integral in nature, such as the minimum pressure on a body or the point of separation on an airfoil; such quantities will be considered in Chap. 5.

To perform an integration over an area or a volume, it is necessary that the integrand be known. The integrand must either be given or information must be available so that it can be approximated with an acceptable degree of accuracy. There are numerous integrands where acceptable approximations cannot be made requiring the solutions of differential equations to provide the required relationships; external flow calculations, such as the lift and drag on an airfoil, often fall into this category. In this chapter, only those problems that involve integral quantities with integrands that are given or that can be approximated will be considered.

4.1 System-to-Control-Volume Transformation

The three basic laws that are of interest in fluid mechanics are often referred to as the conservation of mass, energy, and momentum. The last two are more specifically called the first law of thermodynamics and Newton's second law. Each of these laws is expressed using a Lagrangian description of motion; they apply to a specified mass of the fluid. They are stated as follows:

Mass: The mass of a system remains constant.

Energy: The rate of heat transfer to a system minus the work rate done by a system equals the rate of change of the energy E of the system.

Momentum: The resultant force acting on a system equals the rate of momentum change of the system.

Each of these laws applies to a collection of fluid particles and the density, specific energy, and velocity can vary from point to point in the volume of interest. Using the material derivative and integration over the volume, the laws are now expressed in mathematical terms:

Mass:
$$0 = \frac{D}{Dt} \int_{sys} \rho \, d\mathcal{V} \qquad (4.1)$$

Energy:
$$\dot{Q} - \dot{W} = \frac{D}{Dt} \int_{sys} e\rho \, d\mathcal{V} \qquad (4.2)$$

Momentum:
$$\sum \mathbf{F} = \frac{D}{Dt} \int_{sys} \mathbf{V}\rho \, d\mathcal{V} \qquad (4.3)$$

where the dot over Q and W signifies a time rate and e is the specific energy included in the parentheses of Eq. (1.33). It is very difficult to apply the above equations directly to a collection of fluid particles as the fluid moves along in a simple pipe flow as well as in a more complicated flow, such as flow through a turbine. So, let's convert these integrals that are expressed using a Lagrangian description to integrals expressed using a Eulerian description (see Sec. 3.1). This is a rather tedious

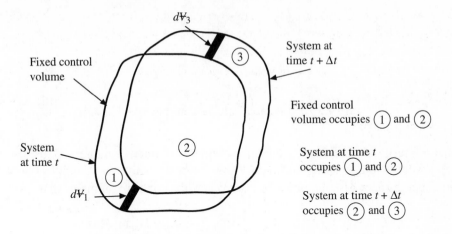

Figure 4.1 The system and the fixed control volume.

derivation but an important one. In this derivation, it is necessary to differentiate between two volumes: a *control volume* that, in this book is a fixed volume in space, and a *system* that is a specified collection of fluid particles. Figure 4.1 illustrates the difference between these two volumes. It represents a general fixed volume in space through which a fluid is flowing; the volumes are shown at time t and at a slightly later time $t + \Delta t$. Let's select the energy $E = \int_{sys} e\rho \, d\forall$ with which to demonstrate the material derivative. We then write, assuming Δt to be a small quantity,

$$
\begin{aligned}
\frac{DE_{sys}}{Dt} &\cong \frac{E_{sys}(t+\Delta t)-E_{sys}(t)}{\Delta t} \\
&= \frac{E_3(t+\Delta t)+E_2(t+\Delta t)-E_1(t)-E_2(t)}{\Delta t} \\
&= \frac{E_2(t+\Delta t)+E_1(t+\Delta t)-E_2(t)-E_1(t)}{\Delta t} + \frac{E_3(t+\Delta t)-E_1(t+\Delta t)}{\Delta t}
\end{aligned}
\tag{4.4}
$$

where we have simply added and subtracted $E_1(t + \Delta t)$ in the last line. Note that the first ratio in the last line above refers to the control volume so that

$$
\frac{E_2(t+\Delta t)+E_1(t+\Delta t)-E_2(t)-E_1(t)}{\Delta t} \cong \frac{dE_{c.v.}}{dt}
\tag{4.5}
$$

where an ordinary derivative is used since we are no longer following a specified fluid mass. Also, we have used "c.v." to denote the control volume. The last ratio in

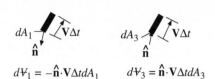

$$dV_1 = -\hat{\mathbf{n}} \cdot \mathbf{V}\Delta t dA_1 \qquad dV_3 = \hat{\mathbf{n}} \cdot \mathbf{V}\Delta t dA_3$$

Figure 4.2 Differential volume elements from Fig. 4.1.

Eq. (4.4) results from fluid flowing into volume 3 and out of volume 1. Consider the differential volumes shown in Fig. 4.1 and displayed with more detail in Fig. 4.2. Note that the area $A_1 + A_3$ completely surrounds the control volume so that

$$E_3(t+\Delta t) - E_1(t+\Delta t) = \int_{A_3} e\,\rho\hat{\mathbf{n}} \cdot \mathbf{V}\Delta t dA_3 + \int_{A_1} e\,\rho\hat{\mathbf{n}} \cdot \mathbf{V}\Delta t dA_1$$

$$= \int_{c.s.} e\,\rho\hat{\mathbf{n}} \cdot \mathbf{V}\Delta t dA \qquad (4.6)$$

where "c.s." is the control surface that surrounds the control volume. Substituting Eqs. (4.5) and (4.6) into Eq. (4.4) results in the *Reynolds transport theorem*, a system-to-control-volume transformation:

$$\frac{DE_{sys}}{Dt} = \frac{d}{dt}\int_{c.v.} e\rho dV + \int_{c.s.} e\rho\hat{\mathbf{n}} \cdot \mathbf{V}dA \qquad (4.7)$$

where, in general, e would represent the specific property of E. Note that we could have taken the limit as $\Delta t \to 0$ to make the derivation more mathematically rigorous.

If we return to the energy equation of Eq. (4.2) we can now write it as

$$\dot{Q} - \dot{W} = \frac{d}{dt}\int_{c.v.} e\rho dV + \int_{c.s.} e\rho\hat{\mathbf{n}} \cdot \mathbf{V}dA \qquad (4.8)$$

If we let $e = 1$ in Eq. (4.7) [see Eq. (4.1)], the conservation of mass results. It is

$$0 = \frac{d}{dt}\int_{c.v.} \rho dV + \int_{c.s.} \rho\hat{\mathbf{n}} \cdot \mathbf{V}dA \qquad (4.9)$$

And finally, if we replace e in Eq. (4.7) with the vector $\mathbf{V}$ [see Eq. (4.3)], Newton's second law results:

$$\Sigma\mathbf{F} = \frac{d}{dt}\int_{c.v.} \rho\mathbf{V}dV + \int_{c.s.} \mathbf{V}\rho\hat{\mathbf{n}} \cdot \mathbf{V}dA \qquad (4.10)$$

These three equations can be written in a slightly different form by recognizing that a *fixed control volume* has been assumed. That means that the limits of the first integral on the right-hand side of each equation are independent of time. Hence, the time derivative can be moved inside the integral if desired. Note that it would be written as a partial derivative should it be moved inside the integral since the integrand depends, in general, on *x, y, z,* and *t.* For example, the momentum equation would take the form

$$\Sigma \mathbf{F} = \int_{\text{c.v.}} \frac{\partial}{\partial t}(\rho \mathbf{V}) d\mathcal{V} + \int_{\text{c.s.}} \mathbf{V} \rho \hat{\mathbf{n}} \cdot \mathbf{V} dA \qquad (4.11)$$

The following three sections will apply these integral forms of the basic laws to problems in which the integrands are given or in which they can be assumed.

4.2 Continuity Equation

The most general relationship for the conservation of mass using the Eulerian description that focuses on a fixed volume was developed in the preceding section as Eq. (4.9). Since the limits on the volume integral do not depend on time, this can be written as

$$0 = \int_{\text{c.v.}} \frac{\partial \rho}{\partial t} d\mathcal{V} + \int_{\text{c.s.}} \rho \hat{\mathbf{n}} \cdot \mathbf{V} dA \qquad (4.12)$$

If the flow of interest can be assumed to be a steady flow so that time does not enter the above equation, the equation simplifies to

$$0 = \int_{\text{c.s.}} \rho \hat{\mathbf{n}} \cdot \mathbf{V} dA \qquad (4.13)$$

Those flows in which the density ρ is uniform over an area are of particular interest in our study of fluids. Also, most applications have one entrance and one exit. For such a problem, the above equation can then be written as

$$\rho_2 A_2 \overline{V}_2 = \rho_1 A_1 \overline{V}_1 \qquad (4.14)$$

where an over bar denotes an average over an area, i.e., $\overline{V} A = \int V dA$. Note also that at an entrance we use $\hat{\mathbf{n}} \cdot \mathbf{V}_1 = -V_1$ since the unit vector points out of the volume, and the velocity is into the volume. But at an exit, $\hat{\mathbf{n}} \cdot \mathbf{V}_2 = V_2$ since the two vectors are in the same direction.

For incompressible flows in which the density does not change[1] between the entrance and the exit, and the velocity is uniform over each area, the conservation of mass takes the simplified form:

$$A_2 V_2 = A_1 V_1 \qquad (4.15)$$

We refer Eqs. (4.12) to (4.15) as the *continuity equation*. These equations are used most often to relate the velocities between sections.

The quantity $\rho A V$ is the *mass flux* and has units of kg/s. The quantity VA is the *flow rate* (or *discharge*) and has units of m^3/s. The mass flux is usually used in a gas flow and the discharge in a liquid flow. They are defined by

$$\dot{m} = \rho A V$$
$$Q = AV \qquad (4.16)$$

where V is the average velocity at a section of the flow.

EXAMPLE 4.1

Water flows in a 6-cm-diameter pipe with a flow rate of 0.02 m^3/s. The pipe is reduced in diameter to 2.8 cm. Calculate the maximum velocity in the pipe. Also calculate the mass flux. Assume uniform velocity profiles.

Solution

The maximum velocity in the pipe will be where the diameter is the smallest. In the 2.8-cm diameter section we have

$$Q = A_2 V_2$$
$$0.02 = \pi \times 0.014^2 V_2 \qquad \therefore V_2 = 32.5 \text{ m/s}$$

The mass flux is

$$\dot{m} = \rho Q = 1000 \times 0.02 = 20 \text{ kg/s}$$

EXAMPLE 4.2

Water flows into a volume that contains a sponge with a flow rate of 0.02 m^3/s. It exits the volume through two tubes, one 2 cm in diameter, and the other with a

---

[1]Not all incompressible flows have constant density. Atmospheric and oceanic flows are examples, as is salt water flowing in a canal.

mass flux of 10 kg/s. If the velocity out of the 2-cm-diameter tube is 15 m/s, determine the rate at which the mass is changing inside the volume.

Solution
The continuity equation (4.9) is used. It is written in the form

$$0 = \frac{dm_{vol}}{dt} + \dot{m}_2 + \rho A_3 V_3 - \rho Q_1$$

where $m_{vol} = \int \rho d V$ and the two exits and entrance account for the other three terms. Expressing the derivative term as $\dot{m}_{vol}$, the continuity equation becomes

$$\dot{m}_{vol} = \rho Q_1 - \dot{m}_2 - \rho A_3 V_3$$
$$= 1000 \times 0.02 - 10 - 1000 \times \pi \times 0.01^2 \times 15 = 5.29 \text{ kg/s}$$

The sponge is soaking up water at the rate of 5.29 kg/s.

4.3 The Energy Equation

The first law of thermodynamics, or simply, the energy equation, is of use whenever heat transfer or work is desired. If there is essentially no heat transfer and no external work from a pump or some other device, the energy equation allows us to relate the pressure, the velocity, and the elevation. We will begin with the energy equation (4.8) in its general form:

$$\dot{Q} - \dot{W} = \frac{d}{dt} \int_{c.v.} e \rho d V + \int_{c.s.} e \rho \hat{n} \cdot V dA \qquad (4.17)$$

Most applications allow a simplified energy equation by assuming a steady, uniform flow with one entrance and one exit, so that

$$\dot{Q} - \dot{W} = e_2 \rho_2 V_2 A_2 - e_1 \rho_1 V_1 A_1 \qquad (4.18)$$

where we have used $\hat{n} \cdot V = -V_1$ at the entrance. Using the continuity equation (4.14), this is written as

$$\dot{Q} - \dot{W} = \dot{m}(e_2 - e_1) \qquad (4.19)$$

The work rate term results from a force moving with a velocity: $\dot{W} = \mathbf{F} \cdot \mathbf{V}$. The force can be a pressure or a shear multiplied by an area. If the flow is in a conduit, e.g., a pipe or a channel, the walls do not move so there is no work done by the walls. If there is a moving belt, there could be an input of work due to the shear between the belt and the fluid. The most common work rate terms result from the pressure forces at the entrance and the exit (pressure is assumed to be uniform over each area) and from any device located between the entrance and the exit. The work rate term is expressed as

$$\dot{W} = p_2 A_2 V_2 - p_1 A_1 V_1 + \dot{W}_S \qquad (4.20)$$

where power output is considered positive and $\dot{W}_S$ is the shaft power output from the control volume (a pump would provide a negative power and a turbine, a positive power output). Using the expression for e given in Eq. (1.33), Eq. (4.19) takes the form

$$\dot{Q} - p_2 A_2 V_2 + p_1 A_1 V_1 - \dot{W}_S = \dot{m}\left(\frac{V_2^2}{2} + gz_2 + \tilde{u}_2 - \frac{V_1^2}{2} - gz_1 - \tilde{u}_1\right) \qquad (4.21)$$

The heat transfer term and the internal energy terms form the losses in the flow (viscous effects result in heat transfer and/or an increase in internal energy). Divide Eq. (4.21) by $\dot{m}g$ and simplify[2]:

$$-\frac{\dot{W}_S}{\dot{m}g} = \frac{V_2^2}{2g} + \frac{p_2}{\gamma_2} + z_2 - \frac{p_1}{\gamma_1} - \frac{V_1^2}{2g} - z_1 + h_L \qquad (4.22)$$

where we have included the loss term as h_L, called the *head loss*; it is

$$h_L = \frac{\tilde{u}_2 - \tilde{u}_1}{g} + \frac{\dot{Q}}{\dot{m}g} \qquad (4.23)$$

The head loss term is often expressed in terms of a *loss coefficient K*:

$$h_L = K\frac{V^2}{2g} \qquad (4.24)$$

---

[2]We used $\dot{m} = \rho_2 A_2 V_2 = \rho_1 A_1 V_1$.

where V is some characteristic velocity in the flow; if it is not obvious it will be specified. Some loss coefficients are listed in Table 7.2; in this chapter they will be given.

The term h_L is called the head loss because it has the dimension of length. We also refer to $V^2/2g$ as the *velocity head*, p/γ as the *pressure head*, and z as the *head*. The sum of these three terms is the *total head*.

An incompressible flow occurs in many applications so that $\gamma_1 = \gamma_2$. Recall that γ for water is 9810 N/m³.

The shaft-work term in Eq. (4.22) is typically due to either a pump or a turbine. If it is a pump, we can define the *pump head H_P* as

$$H_P = \frac{-\dot{W}_S}{\dot{m}g} = \frac{\eta_P \dot{W}_P}{\dot{m}g} \tag{4.25}$$

where $\dot{W}_P$ is the power input to the pump and η_P is the *pump efficiency*. For a turbine the *turbine head H_T* is

$$H_T = \frac{\dot{W}_S}{\dot{m}g} = \frac{\dot{W}_T}{\dot{m}g\eta_T} \tag{4.26}$$

where $\dot{W}_T$ is the power output of the turbine and η_T is the *turbine efficiency*. Power has units of watts or horsepower.

If the flow is not uniform over the entrance and exit, an integration must be performed to obtain the kinetic energy. The rate at which the kinetic energy crosses an area is [see Eqs. (4.17) and (1.33)]

$$\text{Kinetic energy rate} = \int \frac{V^2}{2} \times \rho V dA = \frac{1}{2} \int \rho V^3 dA \tag{4.27}$$

If the velocity distribution is known, the integration can be performed. A *kinetic-energy correction factor* α is defined as

$$\alpha = \frac{\int V^3 dA}{\overline{V}^3 A} \tag{4.28}$$

The kinetic energy term can then be written as

$$\frac{1}{2}\rho \int V^3 dA = \frac{1}{2}\rho \alpha \overline{V}^3 A \tag{4.29}$$

so that, for nonuniform flows, the energy equation takes the form

$$-\frac{\dot{W}_S}{\dot{m}g} = \alpha_2 \frac{\overline{V}_2^2}{2g} + z_2 + \frac{p_2}{\gamma_2} - \alpha_1 \frac{\overline{V}_1^2}{2g} - z_1 - \frac{p_1}{\gamma_1} + h_L \qquad (4.30)$$

where $\overline{V}_1$ and $\overline{V}_2$ are the average velocities at sections 1 and 2, respectively. Equation (4.30) is used if the values for α are known; for parabolic profiles, $\alpha = 2$ in a pipe and $\alpha = 1.5$ between parallel plates. For turbulent flows (most flows in engineering applications), $\alpha \cong 1$.

EXAMPLE 4.3

Water flows from a reservoir with an elevation of 30 m through a 5-cm-diameter pipe that has a 2-cm-diameter nozzle attached to the end, as shown. The loss coefficient for the entire pipe is given as $K = 1.2$. Estimate the flow rate of water through the pipe. Also, predict the pressure just upstream of the nozzle (neglect the losses through the nozzle). The nozzle is at an elevation of 10 m.

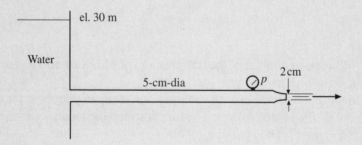

Solution
The energy equation is written in the form

$$-\frac{\cancel{\dot{W}_S}}{\cancel{\dot{m}g}} = \frac{V_2^2}{2g} + z_2 + \frac{\cancel{p_2}}{\cancel{\gamma_2}} - \frac{\cancel{V_1^2}}{\cancel{2g}} - z_1 - \frac{\cancel{p_1}}{\cancel{\gamma_1}} + K\frac{V^2}{2g}$$

where the pressure is zero at surface 1 and at the exit 2, the velocity is zero at the surface, and there is no shaft work (there is no pump or turbine). The loss coefficient would be based on the characteristic velocity V in the pipe, not the exit velocity V_2. Use the continuity equation to relate the velocities:

$$V = \frac{A_2}{A}V_2 = \frac{d_2^2}{d^2}V_2 = \frac{4}{25}V_2$$

The energy equation provides

$$0 = \frac{V_2^2}{2g} + 10 - 30 + 1.2 \times \left(\frac{4}{25}\right)^2 \frac{V_2^2}{2g} \qquad \therefore V_2 = 19.5 \text{ m/s}$$

The pressure just before the nozzle is found by applying the energy equation across the nozzle assuming no losses (Bernoulli's equation could also be used):

$$-\frac{\cancel{\dot{W}_s}}{\dot{m}g} = \frac{V_2^2}{2g} + \frac{\cancel{p_2}}{\cancel{\gamma}} + \cancel{z_2} - \frac{V^2}{2g} - \frac{p}{\gamma} - \cancel{z}$$

where area 2 is at the exit and p and V are upstream of the nozzle. The energy equation gives

$$0 = \frac{19.5^2}{2 \times 9.81} + \frac{\cancel{p_2}}{\cancel{\gamma}} + \cancel{z_2} - \left(\frac{4}{25}\right)^2 \frac{19.5^2}{2 \times 9.8} - \frac{p}{9810} - \cancel{z}$$

$$\therefore p = 185\,300 \text{ Pa} \qquad \text{or} \qquad 185.3 \text{ kPa}$$

EXAMPLE 4.4

An energy conscious couple decides to dam up the creek flowing next to their cabin and estimates that a head of 4 m can be established above the exit to a turbine they bought on eBay. The creek is estimated to have a flow rate of 0.8 m³/s. What is the maximum power output of the turbine assuming no losses and a velocity at the turbine's exit of 3.6 m/s?

Solution
The energy equation is applied as follows:

$$-\frac{\dot{W}_T}{\dot{m}g} = \frac{V_2^2}{2g} + \frac{\cancel{p_2}}{\cancel{\gamma}} + \cancel{z_2} - \frac{\cancel{V_1^2}}{\cancel{2g}} - \frac{\cancel{p_1}}{\cancel{\gamma}} - z_1 + \cancel{h_L}$$

It is only the head of the water above the turbine that provides the power; the exiting velocity subtracts from the power. There results, using $\dot{m} = \rho Q = 1000 \times 0.8 = 800$ kg/s,

$$\dot{W}_T = \dot{m}gz_1 - \dot{m}\frac{V_2^2}{2}$$

$$= 800 \times 9.81 \times 4 - 800 \times \frac{3.6^2}{2} = 26\,200 \text{ J/s} \qquad \text{or} \qquad 26.2 \text{ kW}$$

Let's demonstrate that the units on $\dot{m}gz_1$ are J/s. The units on $\dot{m}gz_1$ are $\dfrac{\text{kg}}{\text{s}} \times \dfrac{\text{m}}{\text{s}^2} \times \text{m} = \dfrac{\text{kg} \cdot \text{m}}{\text{s}^2} \times \dfrac{\text{m}}{\text{s}} = \dfrac{\text{N} \cdot \text{m}}{\text{s}} = \text{J/s}$ where, from $F = ma$, we see that $N = \text{kg} \cdot \text{m/s}^2$. If the proper units are included on the items in our equations, the units will come out as expected, i.e., the units on $\dot{W}_T$ must be J/s.

4.4 The Momentum Equation

When a force is involved in a calculation, it is often necessary to apply Newton's second law, or simply, the *momentum equation*, to the problem of interest. For some general volume, using the Eulerian description of motion, the momentum equation was presented in Eq. (4.10) in its most general form for a fixed control volume as

$$\sum \mathbf{F} = \int_{\text{c.v.}} \frac{\partial(\rho\mathbf{V})}{\partial t}d\mathcal{V} + \int_{\text{c.s.}} \mathbf{V}\rho\hat{\mathbf{n}} \cdot \mathbf{V}dA \qquad (4.31)$$

When applying this equation to a control volume, we must be careful to include all forces acting on the control volume, so it is very important to sketch the control volume and place the forces on the sketch. The control volume takes the place of the free-body diagram utilized in mechanics courses.

Most often, steady, uniform flows with one entrance and one outlet are encountered. For such flows, Eq. (4.31) reduces to

$$\sum \mathbf{F} = \rho_2 A_2 V_2 \mathbf{V}_2 - \rho_1 A_1 V_1 \mathbf{V}_1 \qquad (4.32)$$

Using continuity $\dot{m} = \rho_2 A_2 V_2 = \rho_1 A_1 V_1$, the momentum equation takes the simplified form

$$\sum \mathbf{F} = \dot{m}(\mathbf{V}_2 - \mathbf{V}_1) \qquad (4.33)$$

This is the form most often used when a force is involved in a calculation. It is a vector equation that contains the following three scalar equations (using rectangular coordinates):

$$\begin{aligned}
\sum F_x &= \dot{m}(V_{2x} - V_{1x}) \\
\sum F_y &= \dot{m}(V_{2y} - V_{1y}) \\
\sum F_z &= \dot{m}(V_{2z} - V_{1z})
\end{aligned} \qquad (4.34)$$

If the profiles at the entrance and exit are not uniform, Eq. (4.31) must be used and the integration performed or, if the *momentum-correction factor* β is known, it can be used. It is found from

$$\int_A V^2 dA = \beta \overline{V}^2 A \qquad (4.35)$$

The momentum equation for a steady flow with one entrance and one outlet then takes the form

$$\Sigma \mathbf{F} = \dot{m}(\beta_2 \mathbf{V}_2 - \beta_1 \mathbf{V}_1) \qquad (4.36)$$

where $\mathbf{V}_1$ and $\mathbf{V}_2$ represent the average velocity vectors over the two areas. For parabolic profiles, $\beta = 1.33$ for a pipe and $\beta = 1.2$ for parallel plates. For turbulent flows (most flows in engineering applications), $\beta \cong 1$.

One of the more important applications of the momentum equation is on the deflectors (or vanes) of pumps, turbines, or compressors. The applications involve both stationary deflectors and moving deflectors. The following assumptions are made for both:

- The frictional force between the fluid and the deflector is negligible.
- The pressure is constant as the fluid moves over the deflector.
- The body force is assumed to be negligible.
- The effect of the lateral spreading of the fluid stream is neglected.

A sketch is made of a stationary deflector in Fig. 4.3. Bernoulli's equation predicts that the fluid velocity will not change ($V_2 = V_1$) as the fluid moves over the deflector.

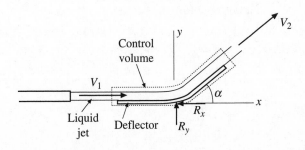

Figure 4.3 A stationary deflector.

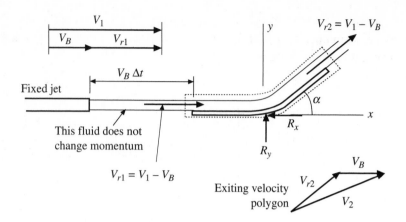

Figure 4.4 A single moving deflector.

Since the pressure does not change, there is no friction, it is a steady flow, and the body forces are neglected. The component momentum equations appear as follows:

$$-R_x = \dot{m}(V_2 \cos\alpha - V_1) = \dot{m}V_1(\cos\alpha - 1)$$
$$R_y = \dot{m}V_2 \sin\alpha = \dot{m}V_1 \sin\alpha \tag{4.37}$$

Given the necessary information, the force components can be calculated.

The analysis of a moving deflector is more complicated. Is it a single deflector (a water scoop to slow a high-speed train), or is it a series of deflectors as in a turbine? First, let us consider a single deflector moving with speed V_B, as shown in Fig. 4.4. The reference frame is attached to the deflector so the flow is steady from such a reference frame[3]. The deflector sees the velocity of the approaching fluid as the relative velocity V_{r1} and it is this relative velocity that Bernoulli's equation predicts will remain constant over the deflector, i.e., $V_{r2} = V_{r1}$. The velocity of the fluid exiting the fixed nozzle is V_1. The momentum equation then provides

$$-R_x = \dot{m}_r(V_1 - V_B)(\cos\alpha - 1)$$
$$R_y = \dot{m}_r(V_1 - V_B)\sin\alpha \tag{4.38}$$

---

[3]If the deflector is observed from the fixed jet, the deflector moves away from the jet and the flow is not a steady flow. It is steady if the flow is observed from the deflector.

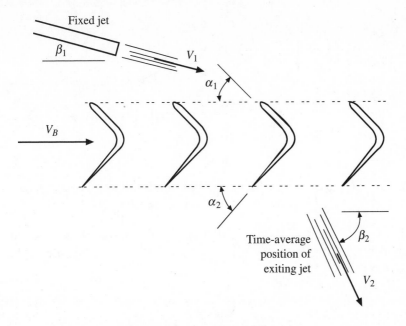

Figure 4.5   A series of vanes.

where $\dot{m}_r$ is that part of the exiting fluid that has its momentum changed. As the deflector moves away from the nozzle, the fluid represented by the length $V_B \Delta t$ does not experience a change in momentum. The mass flux of fluid that experiences a momentum change is

$$\dot{m}_r = \rho A V_{r1} = \rho A (V_1 - V_B) \tag{4.39}$$

which provides us with the relative mass flux used in the expressions for the force components.

For a series of vanes, the nozzles are typically oriented such that the fluid enters the vanes from the side at an angle β_1 and leaves the vanes at an angle β_2, as shown in Fig. 4.5. The vanes are designed so that the relative inlet velocity V_{r1} enters the vanes tangent to a vane (the relative velocity always leaves tangent to the vane) as shown in Fig. 4.6. It is the relative speed that remains constant in magnitude as the fluid moves over the vane, i.e., $V_{r2} = V_{r1}$. We also note that all of the fluid exiting the fixed jet has its momentum changed. So, the expression to determine the x-component of the force is

$$-R_x = \dot{m}(V_{2x} - V_{1x}) \tag{4.40}$$

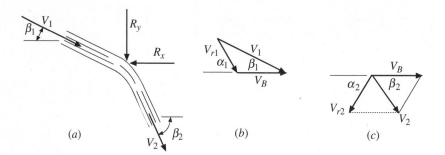

Figure 4.6 (*a*) Average position of the jet, (*b*) the entrance velocity polygon, and (*c*) the exit velocity polygon.

It is this *x*-component of the force that allows the power to be calculated; the *y*-component does no work and hence does not contribute to the power. The power is found from

$$\dot{W} = NR_x V_B \qquad (4.41)$$

where *N* is the number of jets in the device and we have observed that the force R_x moves with velocity V_B.

EXAMPLE 4.5

A 10-cm-diameter hose maintained at a pressure of 1600 kPa provides water from a tanker to a fire. There is a nozzle on the end of the hose that reduces the diameter to 2.5 cm. Estimate the force that the water exerts on the nozzle. The losses can be neglected in a short nozzle.

Solution

A sketch of the water contained in the nozzle is important so that the control volume is carefully identified. It is shown. Note that $p_2 = 0$ and we expect

that the force F_N of the nozzle on the water acts to the left. The velocities are needed upstream and at the exit of the nozzle. Continuity provides

$$A_2 V_2 = A_1 V_1 \qquad \therefore V_2 = \frac{10^2}{2.5^2} V_1 = 16 V_1$$

The energy equation requires

$$\frac{V_2^2}{2}+\frac{\cancel{p_2}}{\rho}+g\cancel{z_2}=\frac{V_1^2}{2}+\frac{p_1}{\rho}+g\cancel{z_1}+\cancel{h_L} \qquad 16^2\frac{V_1^2}{2}=\frac{V_1^2}{2}+\frac{1600\,000}{1000}$$

$$\therefore V_1=3.54 \text{ m/s} \quad \text{and} \quad V_2=56.68 \text{ m/s}$$

The momentum equation then gives

$$p_1A_1-F_N=\dot{m}(V_2-V_1)=\rho A_1V_1(V_2-V_1)=15\rho A_1V_1^2$$

$$1600\,000\times\pi\times0.05^2-F_N=15\times1000\times\pi\times0.05^2\times3.54^2 \qquad \therefore F_N=11090 \text{ N}$$

The force of the water on the nozzle would be equal and opposite to F_N.

EXAMPLE 4.6

A steam turbine contains eight 4-cm-diameter nozzles each accelerating steam to 200 m/s, as shown. The turbine blades are moving at 80 m/s and the density of the steam is 2.2 kg/m³. Calculate the maximum power output.

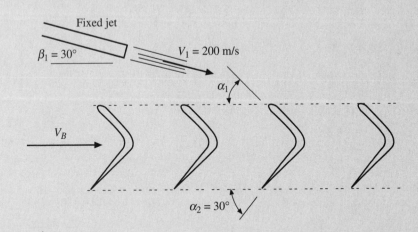

Solution

The angle α_1 is determined from the velocity polygon of Fig. 4.6b. For the x- and y-components, using $V_1 = 200$ m/s and $V_B = 80$ m/s, we have

$$200\sin30° = V_{r1}\sin\alpha_1$$

$$200\cos30° = 80+V_{r1}\cos\alpha_1$$

There are two unknowns in the above two equations: V_{r1} and α_1. A simultaneous solution provides

$$V_{r1} = 136.7 \text{ m/s} \quad \text{and} \quad \alpha_1 = 47.0°$$

Neglecting losses allows $V_{r2} = V_{r1} = 136.7$ m/s so the velocity polygon at the exit (see Fig.4.6c) provides

$$V_2 \sin \beta_2 = 136.7 \sin 30°$$
$$V_2 \cos \beta_2 = 80 - 136.7 \cos 30°$$

These two equations are solved to give

$$V_2 = 78.39 \text{ m/s} \quad \text{and} \quad \beta_2 = 119.3°$$

Observe that the exiting velocity polygon appears as displayed.

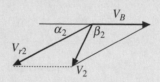

The force acting on the blades due to one nozzle is

$$-F = \dot{m}(V_{2x} - V_{1x})$$
$$= 2.2 \times \pi \times 0.02^2 \times 200(-78.39 \cos 60.7° - 200 \cos 30°) \quad \therefore F = 117 \text{ N}$$

The power output is then

$$\dot{W} = N \times F \times V_B = 8 \times 11.7 \times 80 = 74\,880 \text{ W} \quad \text{or} \quad 100.4 \text{ hp}$$

EXAMPLE 4.7

The relatively rapid flow of water in a horizontal rectangular channel can suddenly "jump" to a higher level (an obstruction downstream may be the cause). This is called a *hydraulic jump*. For the situation shown, calculate the higher depth downstream. Assume uniform flow.

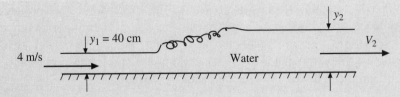

Solution

For a short section of water, the frictional force on the walls can be neglected. The forces acting on the water are F_1 acting to the right and F_2 acting to the left; they are

$$F_1 = \gamma \bar{h}_1 A_1 = 9810 \times 0.20 \times 0.40w = 785w \qquad \text{and} \qquad F_2 = \gamma \bar{h}_2 A_2 = \gamma \frac{y_2}{2} \times y_2 w$$

Applying the momentum equation gives

$$\Sigma F_x = \dot{m}(V_2 - V_1) = \rho A_1 V_1 (V_2 - V_1)$$
$$785w - 4905 \times wy_2^2 = 1000 \times 0.4w \times 4(V_2 - 4)$$

The width w divides out of this equation but there are two unknowns, y_2 and V_2. The continuity equation relates these two variables:

$$A_2 V_2 = A_1 V_1$$
$$wy_2 V_2 = w \times 0.4 \times 4 \qquad \therefore V_2 = \frac{1.6}{y_2}$$

Substitute this into the momentum equation and obtain

$$785 - 4905 y_2^2 = 1600 \left(\frac{1.6}{y_2} - 4 \right)$$

This equation is a cubic but with a little ingenuity it's a quadratic. Let's factor:

$$7^2 (4 - 10 y_2)(4 + 10 y_2) = \frac{1600}{y_2}(1.6 - 4y_2) = \frac{1600}{2.5 y_2}(4 - 10 y_2)$$

The factor $(4 - 10 y_2)$ divides out and a quadratic equation results:

$$y_2^2 + 0.4 y_2 - 1.306 = 0$$

It has two roots. The one of interest is

$$y_2 = 0.96 \text{ m}$$

This rather interesting effect is analogous to the shock wave that occurs in a supersonic gas flow. It is nature's way of moving from something traveling quite fast to something moving relatively slow while maintaining continuity and momentum. The energy that is lost when making this sudden change through the hydraulic jump can be found by using the energy equation.

Quiz No. 1

1. The time derivative can be moved inside the volume integral in the system-to-control-volume transformation because

    (A) The integrand is time independent

    (B) The limits of integration are time independent

    (C) The integral is over space coordinates

    (D) The volume is allowed to deform

2. Air at 25°C and 240 kPa flows in a 10-cm-diameter pipe at 40 m/s. The mass flux is nearest

    (A) 0.94 kg/s

    (B) 1.14 kg/s

    (C) 1.25 kg/s

    (D) 1.67 kg/s

3. Water flows in a 2- by 4-cm rectangular duct at 16 m/s. The duct undergoes a transition to a 6-cm-diameter pipe. Calculate the velocity in the pipe.

    (A) 2.76 m/s

    (B) 3.14 m/s

    (C) 3.95 m/s

    (D) 4.53 m/s

4. A balloon is being filled with water at an instant when the diameter is 50 cm. If the flow rate into the balloon is 0.01 m³/s, the rate of increase in the diameter is nearest

    (A) 2.1 cm/s

    (B) 2.6 cm/s

    (C) 3.2 cm/s

    (D) 3.8 cm/s

5. A sponge is contained in a volume that has one 4-cm-diameter inlet A_1 into which water flows and two 2-cm-diameter outlets, A_2 and A_3. The sponge is to have $dm/dt = 0$. Find V_1 if $Q_2 = 0.002$ m³/s and $\dot{m}_3 = 2.5$ kg/s.

   (A) 3.58 m/s

   (B) 3.94 m/s

   (C) 4.95 m/s

   (D) 5.53 m/s

6. The energy equation $-\dfrac{\dot{W}_S}{\dot{m}g} = \dfrac{V_2^2}{2g} + \dfrac{p_2}{\gamma_2} + z_2 - \dfrac{p_1}{\gamma_1} - \dfrac{V_1^2}{2g} - z_1 + h_L$ does not assume which of the following

   (A) Steady flow

   (B) Incompressible flow

   (C) Uniform flow

   (D) Viscous effects

7. Water enters a horizontal nozzle with diameter $d_1 = 8$ cm at 10 m/s and exits to the atmosphere through a 4-cm-diameter outlet. The pressure upstream of the nozzle is nearest

   (A) 600 kPa

   (B) 650 kPa

   (C) 700 kPa

   (D) 750 kPa

8. Water is transported from one reservoir with surface elevation of 135 m to a lower reservoir with surface elevation of 25 m through a 24-cm-diameter pipe. Estimate the flow rate through the pipe if the loss coefficient between the two surfaces is 20.

   (A) 0.23 m³/s

   (B) 0.34 m³/s

   (C) 0.47 m³/s

   (D) 0.52 m³/s

9. A turbine extracts energy from water flowing through a 10-cm-diameter pipe at a pressure of 800 kPa with an average velocity of 10 m/s. If the turbine is 90 percent efficient, how much energy can be produced if the water is emitted from the turbine to the atmosphere through a 20-cm-diameter pipe?

   (A) 65 kW

   (B) 70 kW

   (C) 75 kW

   (D) 80 kW

10. A 10-cm-diameter hose delivers 0.04 m³/s of water through a 4-cm-diameter nozzle. The force of the water on the nozzle is nearest

   (A) 1065 N

   (B) 1370 N

   (C) 1975 N

   (D) 2780 N

11. A *hydraulic jump* (a sudden jump for no apparent reason) can occur in a rectangular channel with no apparent cause. The momentum equation allows the height downstream to be calculated if the upstream height and velocity are known. Neglect any frictional force on the bottom and sidewalls and determine y_2 in the rectangular channel if $V_1 = 10$ m/s and $y_1 = 50$ cm.

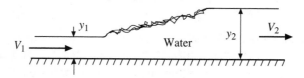

   (A) 2.75 m

   (B) 2.95 m

   (C) 3.15 m

   (D) 3.35 m

12. A 6-cm-diameter horizontal stationary water jet having a velocity of 40 m/s strikes a vertical plate. The force needed to hold the plate if it moves away from the jet at 20 m/s is nearest

   (A) 1365 N

   (B) 1270 N

   (C) 1130 N

   (D) 1080 N

13. The blades of Fig. 4.5 deflect a jet of water having $V_1 = 40$ m/s. Determine the required blade angle α_1 if $\beta_1 = 30°$, $\alpha_2 = 45°$, and $V_B = 20$ m/s.

    (A) 53.8°

    (B) 56.4°

    (C) 58.2°

    (D) 63.4°

14. If the jet in Prob. 13 is 2 cm in diameter, estimate the force of the jet on the blade.

    (A) 387 N

    (B) 404 N

    (C) 487 N

    (D) 521 N

Quiz No. 2

1. Water flows in a 2- by 4-cm rectangular duct at 16 m/s. The duct undergoes a transition to a 6-cm-diameter pipe. Calculate the velocity in the pipe.

2. Air flows in a 20-cm-diameter duct at 120°C and 120 kPa with a mass flux of 5 kg/s. The circular duct converts to a 20-cm square duct in which the temperature and pressure are 140°C and 140 kPa, respectively. Determine the velocity in the square duct.

3. Air at 40°C and 250 kPa is flowing in a 32-cm-diameter pipe at 10 m/s. The pipe changes diameter to 20 cm and the density of the air changes to 3.5 kg/m³. Calculate the velocity in the smaller diameter pipe.

4. Atmospheric air flows over the flat plate as shown. Viscosity makes the air stick to the surface creating a thin boundary layer. Estimate the mass flux $\dot{m}$ of the air across the surface that is 10 cm above the 120-cm-wide plate if $u(y) = 800y$.

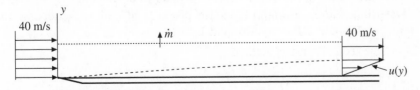

5. A sponge is contained in a volume that has one 4-cm-diameter inlet A_1 into which water flows and two outlets, A_2 and A_3. Determine dm/dt of the sponge if $V_1 = 5$ m/s, $Q_2 = 0.002$ m³/s, and $\dot{m}_3 = 2.5$ kg/s.

6. Water flows from a reservoir with an elevation of 25 m out to a 12-cm-diameter pipe that has a 4-cm-diameter nozzle attached to the end. The loss coefficient for the entire pipe is given as $K = 2$. Estimate the flow rate of water through the pipe. The nozzle is at an elevation of 10 m.

7. A dam is proposed on a remote stream that measures approximately 25-cm deep by 350-cm wide with an average velocity of 2.2 m/s. If the dam can be constructed so that the free surface above a turbine is 10 m, estimate the maximum power output of an 88 percent efficient turbine.

8. An 85 percent efficient pump is used to increase the pressure in water from 120 to 800 kPa in a 10-cm-diameter pipe. What is the required horsepower of the pump for a flow rate of 20 L/s?

9. Air enters a compressor at 25°C and 10 kPa with negligible velocity. It exits through a 2-cm-diameter pipe at 400 kPa and 160°C with a velocity of 200 m/s. Determine the heat transfer if the power required is 18 kW.

10. A turbine is located in a 24-cm-diameter pipe. A piezometer tube upstream of a turbine measures the same pressure as a pitot tube downstream of the turbine. If the upstream velocity is 20 m/s and the turbine is 90 percent efficient, what is the turbine output?

11. A nozzle with exit diameter d is attached to a hose of diameter $3d$ with upstream pressure of 200 kPa. The nozzle changes the direction of the water flow from the hose through an angle of 90°. Calculate the magnitude of the force of the water on the nozzle if $d = 1$ cm.

12. A *hydraulic jump* (a sudden jump for no apparent reason) can occur in a rectangular channel with no apparent cause. The continuity and momentum equations allow the variables to be related. Neglect any frictional force on the bottom and sidewalls and determine V_1 in the rectangular channel if $V_2 = V_1/4$.

13. A 4-cm-diameter horizontal stationary water jet having a velocity of 50 m/s strikes a cone having an included angle at the apex of 60°. The water leaves the cone symmetrically. Determine the force needed to hold the cone if it moves into the jet at 20 m/s.

14. The blades of Fig. 4.6 deflect a 2-cm-diameter jet of water having $V_1 = 40$ m/s. Determine the blade angle α_1 and the power produced by the jet assuming no losses if $\beta_1 = 20°$, $\alpha_2 = 50°$, and $V_B = 15$ m/s.

CHAPTER 5

The Differential Equations

This chapter may be omitted in an introductory course. The derivations in subsequent chapters will either not require these differential equations or there will be two methods to derive the equations: one using differential elements and one utilizing the differential equations.

In Chap. 4 problems were solved using integrals for which the integrands were known or could be approximated. Differential equations are needed in order to solve for those quantities in the integrands that are not known, such as the velocity distribution in a pipe or the velocity and pressure distributions on and around an airfoil. The differential equations may also contain information of interest, such as a point of separation of a fluid from a surface.

5.1 The Boundary-Value Problem

To solve a partial differential equation for the dependent variable, certain conditions are required, i.e., the dependent variable must be specified at certain values of the independent variables. If the independent variables are space coordinates (such as the velocity at the wall of a pipe), the conditions are called *boundary conditions*. If the independent variable is time, the conditions are called *initial conditions*. The general problem is usually referred to as a *boundary-value problem*.

Boundary conditions typically result from one or more of the following:

- The no-slip condition in a viscous flow. Viscosity causes any fluid, be it a gas or a liquid, to stick to the boundary. Most often the boundary is not moving.

- The normal component of the velocity in an inviscid flow. In an inviscid flow where the viscosity is neglected, the velocity vector is tangent to the boundary at the boundary, providing the boundary is not porous.

- The pressure at a free surface. For problems involving a free surface, a pressure condition is known at the free surface.

For an unsteady flow, initial conditions are required, e.g., the initial velocity and pressure must be specified at some time, usually at $t = 0$.

The differential equations in this chapter will be derived using rectangular coordinates. It may be easier to solve problems using cylindrical or spherical coordinates, so the differential equations using those two coordinate systems will be presented in Table 5.1.

The differential energy equation will not be derived in this book. It would be needed if there are temperature differences on the boundaries or if viscous effects are so large that temperature gradients are developed in the flow. A course in heat transfer would include such effects.

5.2 The Differential Continuity Equation

To derive the differential continuity equation, the infinitesimal element of Fig. 5.1 is needed. It is a small control volume into and from which the fluid flows. It is shown in the xy-plane with depth dz. Let us assume that the flow is only in the xy-plane so that no fluid flows in the z-direction. Since mass could be changing inside the element, the mass that flows into the element minus that which flows out must equal the change in mass inside the element. This is expressed as

Table 5.1   The Differential Continuity, Momentum Equations, and Stresses for Incompressible Flows in Cylindrical and Spherical Coordinates

Continuity
Cylindrical
$$\frac{1}{r}\frac{\partial}{\partial r}(rv_r) + \frac{1}{r}\frac{\partial v_\theta}{\partial \theta} + \frac{\partial v_z}{\partial z} = 0$$
Spherical
$$\frac{1}{r^2}\frac{\partial}{\partial r}(r^2 v_r) + \frac{1}{r\sin\theta}\frac{\partial}{\partial \theta}(v_\theta \sin\theta) + \frac{1}{r\sin\theta}\frac{\partial v_\phi}{\partial \phi} = 0$$
Momentum
Cylindrical
$$\rho\frac{Dv_r}{Dt} - \frac{v_\theta^2}{r} = -\frac{\partial p}{\partial r} + \rho g_r + \mu\left(\nabla^2 v_r - \frac{v_r}{r^2} - \frac{2}{r^2}\frac{\partial v_\theta}{\partial \theta}\right)$$ $$\rho\frac{Dv_\theta}{Dt} + \frac{v_\theta v_r}{r} = -\frac{1}{r}\frac{\partial p}{\partial \theta} + \rho g_\theta + \mu\left(\nabla^2 v_\theta - \frac{v_\theta}{r^2} + \frac{2}{r^2}\frac{\partial v_r}{\partial \theta}\right)$$ $$\rho\frac{Dv_z}{Dt} = -\frac{\partial p}{\partial z} + \rho g_z + \mu\nabla^2 v_z$$
where
$$\frac{D}{Dt} = v_r\frac{\partial}{\partial r} + \frac{v_\theta}{r}\frac{\partial}{\partial \theta} + v_z\frac{\partial}{\partial z} + \frac{\partial}{\partial t}$$ $$\nabla^2 = \frac{\partial^2}{\partial r^2} + \frac{1}{r}\frac{\partial}{\partial r} + \frac{1}{r^2}\frac{\partial^2}{\partial \theta^2} + \frac{\partial^2}{\partial z^2}$$
Spherical
$$\rho\frac{Dv_r}{Dt} - \rho\frac{v_\theta^2 + v_\phi^2}{r} = -\frac{\partial p}{\partial r} + \rho g_r$$ $$+ \mu\left(\nabla^2 v_r - \frac{2v_r}{r^2} - \frac{2}{r^2}\frac{\partial v_\theta}{\partial \theta} - \frac{2v_\theta\cot\theta}{r^2} - \frac{2}{r^2\sin\theta}\frac{\partial v_\phi}{\partial \phi}\right)$$ $$\rho\frac{Dv_\theta}{Dt} + \rho\frac{v_r v_\theta + v_\phi^2\cot\theta}{r} = -\frac{1}{r}\frac{\partial p}{\partial \theta} + \rho g_\theta$$ $$+ \mu\left(\nabla^2 v_\theta + \frac{2}{r^2}\frac{\partial v_r}{\partial \theta} - \frac{v_\theta}{r^2\sin\theta} - \frac{2\cos\theta}{r^2\sin^2\theta}\frac{\partial v_\phi}{\partial \phi}\right)$$

(Continued)

Table 5.1 The Differential Continuity, Momentum Equations, and Stresses for Incompressible Flows in Cylindrical and Spherical Coordinates (*Continued*)

$$\rho\frac{Dv_\phi}{Dt} - \rho\frac{v_r v_\phi + v_\theta v_\phi \cot\theta}{r} = -\frac{1}{r\sin\theta}\frac{\partial p}{\partial\phi} + \rho g_\phi$$

$$+ \mu\left(\nabla^2 v_\phi - \frac{v_\phi}{r^2\sin^2\theta} + \frac{2}{r^2\sin^2\theta}\frac{\partial v_r}{\partial\phi} + \frac{2\cos\theta}{r^2\sin^2\theta}\frac{\partial v_\theta}{\partial\phi}\right)$$

where

$$\frac{D}{Dt} = v_r\frac{\partial}{\partial r} + \frac{v_\theta}{r}\frac{\partial}{\partial\theta} + \frac{v_\phi}{r\sin\theta}\frac{\partial}{\partial\phi} + \frac{\partial}{\partial t}$$

$$\nabla^2 = \frac{1}{r^2}\frac{\partial}{\partial r}\left(r^2\frac{\partial}{\partial r}\right) + \frac{1}{r^2\sin\theta}\frac{\partial}{\partial\theta}\left(\sin\theta\frac{\partial}{\partial\theta}\right) + \frac{1}{r^2\sin\theta}\frac{\partial^2}{\partial\phi^2}$$

Stresses

Cylindrical

$$\sigma_{rr} = -p + 2\mu\frac{\partial v_r}{\partial y} \qquad \tau_{r\theta} = \mu\left(r\frac{\partial(v_\theta/r)}{\partial r} + \frac{1}{r}\frac{\partial v_z}{\partial\theta}\right)$$

$$\sigma_{\theta\theta} = -p + 2\mu\left(\frac{1}{r}\frac{\partial v_\theta}{\partial\theta} + \frac{v_r}{r}\right) \qquad \tau_{\theta z} = \mu\left(\frac{\partial v_\theta}{\partial z} + \frac{1}{r}\frac{\partial v_r}{\partial\theta}\right)$$

$$\sigma_{zz} = -p + 2\mu\frac{\partial v_z}{\partial z} \qquad \tau_{rz} = \mu\left(\frac{\partial v_r}{\partial z} + \frac{\partial v_z}{\partial r}\right)$$

Spherical

$$\sigma_{rr} = -p + 2\mu\frac{\partial v_r}{\partial r} \qquad \tau_{r\theta} = \mu\left[r\frac{\partial}{\partial r}\left(\frac{v_\theta}{r}\right) + \frac{1}{r}\frac{\partial v_r}{\partial\theta}\right]$$

$$\sigma_{\theta\theta} = -p + 2\mu\left(\frac{1}{r\sin\theta}\frac{\partial v_\theta}{\partial\theta} + \frac{v_r}{r}\right) \qquad \tau_{\theta\phi} = \mu\left[\frac{\sin\theta}{r}\frac{\partial}{\partial\theta}\left(\frac{v_\phi}{\sin\theta}\right) + \frac{1}{r\sin\theta}\frac{\partial v_\theta}{\partial\phi}\right]$$

$$\sigma_{\phi\phi} = -p + 2\mu\left(\frac{1}{r\sin\theta}\frac{\partial v_\phi}{\partial\phi} + \frac{v_r}{r} + \frac{v_\theta\cot\theta}{r}\right) \qquad \tau_{r\phi} = \mu\left[\frac{1}{r\sin\theta}\frac{\partial v_r}{\partial\phi} + r\frac{\partial}{\partial r}\left(\frac{v_\phi}{r}\right)\right]$$

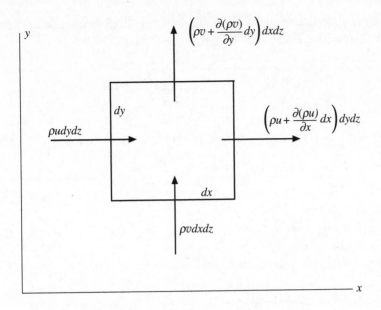

Figure 5.1   Infinitesimal control volume.

$$\rho u \, dy dz - \left(\rho u + \frac{\partial(\rho u)}{\partial x} dx \right) dy dz + \rho v \, dx dz$$

$$- \left(\rho v + \frac{\partial(\rho v)}{\partial y} dy \right) dx dz = \frac{\partial}{\partial t}(\rho \, dx dy dz) \qquad (5.1)$$

where the products ρu and ρv are allowed to change across the element.[1] Simplifying the above, recognizing that the elemental control volume is fixed, results in

$$\frac{\partial(\rho u)}{\partial x} + \frac{\partial(\rho v)}{\partial y} = -\frac{\partial \rho}{\partial t} \qquad (5.2)$$

---

[1]The product could have been included as $\left(\rho + \frac{\partial \rho}{\partial x} dx \right)\left(u + \frac{\partial u}{\partial x} dx \right) dy dz$ on the right-hand side of the element but the above simpler expression is equivalent.

Differentiate the products and include the variation in the z-direction. Then the differential continuity equation can be put in the form

$$\frac{\partial \rho}{\partial t} + u\frac{\partial \rho}{\partial x} + v\frac{\partial \rho}{\partial y} + w\frac{\partial \rho}{\partial z} + \rho\left(\frac{\partial u}{\partial x} + \frac{\partial v}{\partial y} + \frac{\partial w}{\partial z}\right) = 0 \qquad (5.3)$$

The first four terms form the material derivative [see Eq. (3.11)], so Eq. (5.3) becomes

$$\frac{D\rho}{Dt} + \rho\left(\frac{\partial u}{\partial x} + \frac{\partial v}{\partial y} + \frac{\partial w}{\partial z}\right) = 0 \qquad (5.4)$$

providing the most general form of the *differential continuity equation* expressed using rectangular coordinates.

The differential continuity equation is often written using the vector operator

$$\nabla = \frac{\partial}{\partial x}\mathbf{i} + \frac{\partial}{\partial y}\mathbf{j} + \frac{\partial}{\partial z}\mathbf{k} \qquad (5.5)$$

so that Eq. (5.4) takes the form

$$\frac{D\rho}{Dt} + \rho\nabla\cdot\mathbf{V} = 0 \qquad (5.6)$$

where the velocity vector is $\mathbf{V} = u\mathbf{i} + v\mathbf{j} + w\mathbf{k}$. The scalar $\nabla\cdot\mathbf{V}$ is called the *divergence* of the velocity vector.

For an incompressible flow, the density of a fluid particle remains constant as it travels through a flow field, that is,

$$\frac{D\rho}{Dt} = \frac{\partial \rho}{\partial t} + u\frac{\partial \rho}{\partial x} + v\frac{\partial \rho}{\partial y} + w\frac{\partial \rho}{\partial z} = 0 \qquad (5.7)$$

so it is not necessary that the density be constant. If the density is constant, as it often is, then each term in Eq. (5.7) is zero. For an incompressible flow, Eqs. (5.4) and (5.6) also demand that

$$\frac{\partial u}{\partial x} + \frac{\partial v}{\partial y} + \frac{\partial w}{\partial z} = 0 \qquad \text{or} \qquad \nabla\cdot\mathbf{V} = 0 \qquad (5.8)$$

The differential continuity equation for an incompressible flow is presented in cylindrical and spherical coordinates in Table 5.1.

EXAMPLE 5.1
Air flows with a uniform velocity in a pipe with the velocities measured along the centerline at 40-cm increments as shown. If the density at point 2 is 1.2 kg/m³, estimate the density gradient at point 2.

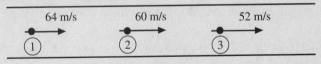

Solution
The continuity equation (5.3) is used since the density is changing. It is simplified as follows:

$$\frac{\partial \rlap{/}\rho}{\partial \rlap{/}t} + u\frac{\partial \rho}{\partial x} + v\frac{\partial \rlap{/}\rho}{\partial \rlap{/}y} + w\frac{\partial \rlap{/}\rho}{\partial \rlap{/}z} + \rho\left(\frac{\partial u}{\partial x} + \frac{\partial \rlap{/}v}{\partial \rlap{/}y} + \frac{\partial \rlap{/}w}{\partial \rlap{/}z}\right) = 0 \qquad \therefore u\frac{\partial \rho}{\partial x} = -\rho\frac{\partial u}{\partial x}$$

Central differences[2] are used to approximate the velocity gradient $\partial u/\partial x$ at point 2 since information at three points is given:

$$\frac{\partial u}{\partial x} \cong \frac{\Delta u}{\Delta x} = \frac{52-64}{0.80} = -15 \; \frac{\text{m/s}}{\text{m}}$$

The best estimate of the density gradient, using the information given, is then

$$\frac{\partial \rho}{\partial x} = -\frac{\rho}{u}\frac{\partial u}{\partial x} = -\frac{1.2}{60}(-15) = 0.3 \; \text{kg/m}^4$$

5.3 The Navier-Stokes Equations

The differential continuity equation derived in Sec. 5.2 contains the three velocity components as the dependent variables for an incompressible flow. If there is a flow of interest in which the velocity field and pressure field are not known, such as the flow around a turbine blade or over a weir, the differential momentum equation

---

[2]A forward difference would give $\partial u/\partial x \cong (52-60)/0.40 = -20$. A backward difference would provide $\partial u/\partial x \cong (60-64)/0.40 = -10$. The central difference is the best linear approximation.

provides three additional equations since it is a vector equation containing three component equations. The four unknowns are then u, v, w, and p when using a rectangular coordinate system. The four equations provide us with the necessary equations and initial and boundary conditions allow a tractable problem. The problems of the turbine blade and the weir are quite difficult to solve and their solutions will not be attempted in this book. Our focus will be solving problems with simple geometries.

Now we will derive the differential momentum equation, a rather challenging task. First, stresses exist on the faces of an infinitesimal, rectangular fluid element, as shown in Fig. 5.2 for the xy-plane. Similar stress components act in the z-direction. The *normal stresses* are designated with σ and the *shear stresses* with τ. There are nine stress components: $\sigma_{xx}, \sigma_{yy}, \sigma_{zz}, \tau_{xy}, \tau_{yx}, \tau_{xz}, \tau_{zx}, \tau_{yz},$ and τ_{zy}. If moments are taken about the x-axis, the y-axis, and the z-axis, respectively, they would show that

$$\tau_{yx} = \tau_{xy} \qquad \tau_{zx} = \tau_{xz} \qquad \tau_{zy} = \tau_{yz} \tag{5.9}$$

So, there are six stress components that must be related to the pressure and velocity components. Such relationships are called *constitutive equations*; they are equations that are not derived but are found using observations in the laboratory.

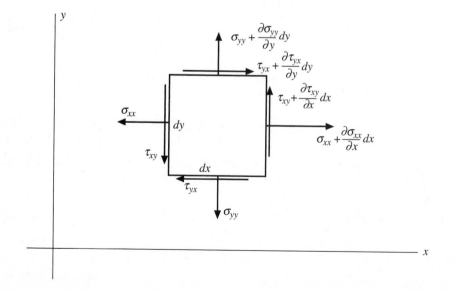

Figure 5.2   Rectangular stress components on a fluid element.

Next, apply Newton's second law to the element of Fig. 5.2, assuming no shear stresses act in the z-direction (we'll simply add those in later) and that gravity acts in the z-direction only:

$$\left(\sigma_{xx}+\frac{\partial\sigma_{xx}}{\partial x}dx\right)dydz-\sigma_{xx}dydz+\left(\tau_{xy}+\frac{\partial\tau_{xy}}{\partial y}dy\right)dxdz-\tau_{xy}dxdz=\rho dxdydz\frac{Du}{Dt}$$

$$\left(\sigma_{yy}+\frac{\partial\sigma_{yy}}{\partial y}dy\right)dxdz-\sigma_{yy}dxdz+\left(\tau_{xy}+\frac{\partial\tau_{xy}}{\partial x}dx\right)dydz-\tau_{xy}dydz=\rho dxdydz\frac{Dv}{Dt}$$

$$(5.10)$$

These are simplified to

$$\frac{\partial\sigma_{xx}}{\partial x}+\frac{\partial\tau_{xy}}{\partial y}=\rho\frac{Du}{Dt}$$

$$(5.11)$$

$$\frac{\partial\sigma_{yy}}{\partial y}+\frac{\partial\tau_{xy}}{\partial x}=\rho\frac{Dv}{Dt}$$

If the z-direction components are included, the differential equations become

$$\frac{\partial\sigma_{xx}}{\partial x}+\frac{\partial\tau_{xy}}{\partial y}+\frac{\partial\tau_{xz}}{\partial z}=\rho\frac{Du}{Dt}$$

$$\frac{\partial\sigma_{yy}}{\partial y}+\frac{\partial\tau_{xy}}{\partial x}+\frac{\partial\tau_{yz}}{\partial z}=\rho\frac{Dv}{Dt}$$

$$(5.12)$$

$$\frac{\partial\sigma_{zz}}{\partial z}+\frac{\partial\tau_{xz}}{\partial x}+\frac{\partial\tau_{yz}}{\partial y}-\rho g=\rho\frac{Dw}{Dt}$$

assuming the gravity term $\rho gdxdydz$ acts in the negative z-direction.

In many flows, the viscous effects that lead to the shear stresses can be neglected and the normal stresses are the negative of the pressure. For such inviscid flows, Eq. (5.12) takes the form

$$\rho\frac{Du}{Dt}=-\frac{\partial p}{\partial x}\qquad\rho\frac{Dv}{Dt}=-\frac{\partial p}{\partial y}\qquad\rho\frac{Dw}{Dt}=-\frac{\partial p}{\partial z}-\rho g\qquad(5.13)$$

In vector form [see Eq. (5.5)], they become the famous *Euler's equation*,

$$\rho\frac{D\mathbf{V}}{Dt}=-\nabla p-\rho g\hat{\mathbf{k}}\qquad(5.14)$$

which is applicable to inviscid flows. For a constant-density, steady flow, Eq. (5.14) can be integrated along a streamline to provide Bernoulli's equation [Eq. (3.27)].

If viscosity significantly effects the flow, Eq. (5.12) must be used. Constitutive equations[3] relate the stresses to the velocity and pressure fields. For a Newtonian[4], isotropic[5] fluid, they have been observed to be

$$\sigma_{xx} = -p + 2\mu\frac{\partial u}{\partial x} + \lambda \mathbf{V} \cdot \mathbf{V} \qquad \tau_{xy} = \mu\left(\frac{\partial u}{\partial y} + \frac{\partial v}{\partial x}\right)$$

$$\sigma_{yy} = -p + 2\mu\frac{\partial v}{\partial y} + \lambda \mathbf{V} \cdot \mathbf{V} \qquad \tau_{xz} = \mu\left(\frac{\partial u}{\partial z} + \frac{\partial w}{\partial x}\right) \qquad (5.15)$$

$$\sigma_{zz} = -p + 2\mu\frac{\partial w}{\partial z} + \lambda \mathbf{V} \cdot \mathbf{V} \qquad \tau_{yz} = \mu\left(\frac{\partial v}{\partial z} + \frac{\partial w}{\partial y}\right)$$

For most gases, *Stokes hypothesis* can be used: $\lambda = -2\mu/3$. If the above normal stresses are added, there results

$$p = -\frac{1}{3}(\sigma_{xx} + \sigma_{yy} + \sigma_{zz}) \qquad (5.16)$$

showing that the pressure is the negative average of the three normal stresses in most gases, including air, and in all liquids in which $\mathbf{V} \cdot \mathbf{V} = 0$.

If Eq. (5.15) is substituted into Eq. (5.12) using $\lambda = -2\mu/3$, there results

$$\rho\frac{Du}{Dt} = -\frac{\partial p}{\partial x} + \mu\left(\frac{\partial^2 u}{\partial x^2} + \frac{\partial^2 u}{\partial y^2} + \frac{\partial^2 u}{\partial z^2}\right) + \frac{\mu}{3}\frac{\partial}{\partial x}\left(\frac{\partial u}{\partial x} + \frac{\partial v}{\partial y} + \frac{\partial w}{\partial z}\right)$$

$$\rho\frac{Dv}{Dt} = -\frac{\partial p}{\partial y} + \mu\left(\frac{\partial^2 v}{\partial x^2} + \frac{\partial^2 v}{\partial y^2} + \frac{\partial^2 v}{\partial z^2}\right) + \frac{\mu}{3}\frac{\partial}{\partial y}\left(\frac{\partial u}{\partial x} + \frac{\partial v}{\partial y} + \frac{\partial w}{\partial z}\right) \qquad (5.17)$$

$$\rho\frac{Dw}{Dt} = -\frac{\partial p}{\partial z} + \mu\left(\frac{\partial^2 w}{\partial x^2} + \frac{\partial^2 w}{\partial y^2} + \frac{\partial^2 w}{\partial z^2}\right) + \frac{\mu}{3}\frac{\partial}{\partial z}\left(\frac{\partial u}{\partial x} + \frac{\partial v}{\partial y} + \frac{\partial w}{\partial z}\right) - \rho g$$

---

[3]The constitutive equations for cylindrical and spherical coordinates are displayed in Table 5.1.

[4]A Newtonian fluid has a linear stress-strain rate relationship.

[5]An isotropic fluid has properties that are independent of direction at a point.

where gravity acts in the negative z-direction and a *homogeneous fluid*[6] has been assumed so that, e.g., $\partial\mu/\partial x = 0$.

Finally, if an incompressible flow is assumed so that $\nabla \cdot \mathbf{V} = 0$, the *Navier-Stokes equations* result:

$$\rho\frac{Du}{Dt} = -\frac{\partial p}{\partial x} + \mu\left(\frac{\partial^2 u}{\partial x^2} + \frac{\partial^2 u}{\partial y^2} + \frac{\partial^2 u}{\partial z^2}\right)$$

$$\rho\frac{Dv}{Dt} = -\frac{\partial p}{\partial y} + \mu\left(\frac{\partial^2 v}{\partial x^2} + \frac{\partial^2 v}{\partial y^2} + \frac{\partial^2 v}{\partial z^2}\right) \tag{5.18}$$

$$\rho\frac{Dw}{Dt} = -\frac{\partial p}{\partial z} + \mu\left(\frac{\partial^2 w}{\partial x^2} + \frac{\partial^2 w}{\partial y^2} + \frac{\partial^2 w}{\partial z^2}\right) - \rho g$$

where the z-direction is vertical. If we introduce the scalar operator called the *Laplacian*, defined by

$$\nabla^2 = \frac{\partial^2}{\partial x^2} + \frac{\partial^2}{\partial y^2} + \frac{\partial^2}{\partial z^2} \tag{5.19}$$

and review the steps leading from Eq. (5.13) to Eq. (5.14), the Navier-Stokes equations can be written in vector form as

$$\rho\frac{D\mathbf{V}}{Dt} = -\nabla p + \mu\nabla^2\mathbf{V} + \rho\mathbf{g} \tag{5.20}$$

The Navier-Stokes equations expressed in cylindrical and spherical coordinates are presented in Table 5.1.

The three scalar Navier-Stokes equations and the continuity equation constitute the four equations that can be used to find the four variables u, v, w, and p provided there are appropriate initial and boundary conditions. The equations are nonlinear due to the acceleration terms, such as $u\partial u/\partial x$ on the left-hand side; consequently, the solution to these equation may not be unique. For example, the flow between two rotating cylinders can be solved using the Navier-Stokes equations to be a relatively simple flow with circular streamlines; it could also be a flow with streamlines that are like a spring wound around the cylinders as a torus; and, there are even

---

[6]A homogeneous fluid has properties that are independent of position.

more complex flows that are also solutions to the Navier-Stokes equations, all satisfying the identical boundary conditions.

The Navier-Stokes equations can be solved with relative ease for some simple geometries. But, the equations cannot be solved for a turbulent flow even for the simplest of examples; a turbulent flow is highly unsteady and three-dimensional and thus requires that the three velocity components be specified at all points in a region of interest at some initial time, say $t = 0$. Such information would be nearly impossible to obtain, even for the simplest geometry. Consequently, the solutions of turbulent flows are left to the experimentalist and are not attempted by solving the equations.

EXAMPLE 5.2
Water flows from a reservoir in between two closely aligned parallel plates, as shown. Write the simplified equations needed to find the steady-state velocity and pressure distributions between the two plates. Neglect any z-variation of the distributions and any gravity effects. Do not neglect $v(x, y)$.

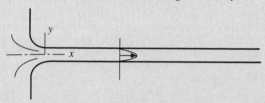

Solution
The continuity equation is simplified, for the incompressible water flow, to

$$\frac{\partial u}{\partial x} + \frac{\partial v}{\partial y} + \cancel{\frac{\partial w}{\partial z}} = 0$$

The differential momentum equations, recognizing that

$$\frac{D}{Dt} = u\frac{\partial}{\partial x} + v\frac{\partial}{\partial y} + \cancel{w\frac{\partial}{\partial z}} + \cancel{\frac{\partial}{\partial t}}$$

are simplified as follows:

$$\rho\left(u\frac{\partial u}{\partial x} + v\frac{\partial u}{\partial y}\right) = -\frac{\partial p}{\partial x} + \mu\left(\frac{\partial^2 u}{\partial x^2} + \frac{\partial^2 u}{\partial y^2} + \cancel{\frac{\partial^2 u}{\partial z^2}}\right)$$

$$\rho\left(u\frac{\partial v}{\partial x} + v\frac{\partial v}{\partial y}\right) = -\cancel{\frac{\partial p}{\partial y}} + \mu\left(\frac{\partial^2 v}{\partial x^2} + \frac{\partial^2 v}{\partial y^2} + \cancel{\frac{\partial^2 p}{\partial z^2}}\right)$$

neglecting pressure variation in the y-direction since the plates are assumed to be a relatively small distance apart. So, the three equations that contain the three variables u, v, and p are

$$\frac{\partial u}{\partial x} + \frac{\partial v}{\partial y} = 0$$

$$\rho\left(u\frac{\partial u}{\partial x} + v\frac{\partial u}{\partial y}\right) = -\frac{\partial p}{\partial x} + \mu\left(\frac{\partial^2 u}{\partial x^2} + \frac{\partial^2 u}{\partial y^2}\right)$$

$$\rho\left(u\frac{\partial v}{\partial x} + v\frac{\partial v}{\partial y}\right) = \mu\left(\frac{\partial^2 v}{\partial x^2} + \frac{\partial^2 v}{\partial y^2}\right)$$

To find a solution to these equations for the three variables, it would be necessary to use the no-slip conditions on the two plates and assumed boundary conditions at the entrance, which would include $u(0, y)$ and $v(0, y)$. Even for this rather simple geometry, the solution to this entrance-flow problem appears, and is, quite difficult. A numerical solution could be attempted.

EXAMPLE 5.3

Integrate Euler's equation [Eq. (5.14)] along a streamline as shown for a steady, constant-density flow and show that Bernoulli's equation (3.25) results.

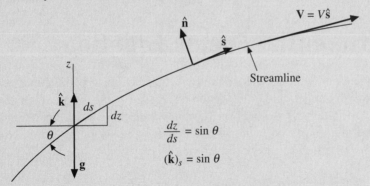

Solution

First, sketch a general streamline and show the selected coordinates normal to and along the streamline so that the velocity vector can be written as $V\hat{s}$, as we did in Fig. 3.10. First, express DV/dt in these coordinates:

$$\frac{D\mathbf{V}}{Dt} = \cancel{\frac{\partial \mathbf{V}}{\partial t}} + V\frac{\partial(V\hat{s})}{\partial s} + \cancel{\mathbf{V}_n}\frac{\partial \mathbf{V}}{\partial n} = V\hat{s}\frac{\partial V}{\partial s} + V^2\frac{\partial\hat{s}}{\partial s}$$

where $\partial\hat{s}/\partial s$ is nonzero since $\hat{s}$ can change direction from point to point on the streamline; it is a vector quantity in the $\hat{n}$ direction. Applying Euler's equation along a streamline (in the s-direction) allows us to write

$$\rho V \frac{\partial V}{\partial s} = -\frac{\partial p}{\partial s} - \rho g \frac{\partial z}{\partial s}$$

where we write $(\hat{k})_s = \partial z/\partial s$. Partial derivatives are necessary because quantities can vary in the normal direction. The above equation is then written as

$$\frac{\partial}{\partial s}\left(\rho \frac{V^2}{2} + p + \rho gz\right) = 0$$

provided the density ρ is constant. This means that along a streamline,

$$\frac{V^2}{2} + \frac{p}{\rho} + gz = \text{const}$$

This is Bernoulli's equation; it requires the same conditions as it did when it was derived in Chap. 3.

5.4 The Differential Energy Equation

Most problems in an introductory fluid mechanics course involve isothermal fluid flows in which temperature gradients do not exist. So, the differential energy equation is not of interest. The study of flows in which there are temperature gradients is included in a course on heat transfer. For completeness, the differential energy equation is presented here without derivation. In general, it is

$$\rho \frac{Dh}{Dt} = K\nabla^2 T + \frac{Dp}{Dt} \tag{5.21}$$

where K is the *thermal conductivity*. For an incompressible ideal gas flow it becomes

$$\rho c_p \frac{DT}{Dt} = K\nabla^2 T \tag{5.22}$$

For a liquid flow it takes the form

$$\frac{DT}{Dt} = \alpha \nabla^2 T \tag{5.23}$$

where α is the *thermal diffusivity* defined by $\alpha = K/\rho c_p$.

Quiz No. 1

1. The x-component of the velocity in a certain plane flow depends only on
   y by the relationship $u(y) = Ay$. Determine the y-component $v(x, y)$ of the
   velocity if $v(x, 0) = 0$.

   (A) 0

   (B) Ax

   (C) Ay

   (D) Axy

2. If $u =$ Const in a plane flow, what can be said about $v(x, y)$?

   (A) 0

   (B) $f(x)$

   (C) $f(y)$

   (D) $f(x, y)$

3. If, in a plane flow, the two velocity components are given by $u(x,y) =$
   $8(x^2 + y^2)$ and $v(x, y) = 8xy$. What is $D\rho/Dt$ at $(1, 2)$ m if at that point
   $\rho = 2$ kg/m³?

   (A) -24 kg/m³/s

   (B) -32 kg/m³/s

   (C) -48 kg/m³/s

   (D) -64 kg/m³/s

4. If $u(x, y) = 4 + \dfrac{2x}{x^2 + y^2}$ in a plane incompressible flow, what is $v(x, y)$ if
   $v(x, 0) = 0$?

   (A) $2y/(x^2 + y^2)$

   (B) $-2x/(x^2 + y^2)$

   (C) $-2y/(x^2 + y^2)$

   (D) $2x/(x^2 + y^2)$

5. The velocity component $v_\theta = -(25 + 1/r^2)\cos\theta$ in a plane incompressible flow. Find $v_r(r, \theta)$ if $v_r(r, 0) = 0$.

    (A) $25(1 - 1/r^2)\sin\theta$

    (B) $(25 - 1/r^2)\sin\theta$

    (C) $-25(1 - 1/r^2)\sin\theta$

    (D) $-(25 - 1/r^2)\sin\theta$

6. Simplify the appropriate Navier-Stokes equation for the flow between parallel plates assuming $u = u(y)$ and gravity in the z-direction. The streamlines are assumed to be parallel to the plates so that $v = w = 0$.

    (A) $\partial p/\partial x = \mu \partial^2 u/\partial y^2$

    (B) $\partial p/\partial y = \mu \partial^2 u/\partial x^2$

    (C) $\partial p/\partial y = \mu \partial^2 u/\partial y^2$

    (D) $\partial p/\partial x = \mu \partial^2 u/\partial x^2$

Quiz No. 2

1. A compressible flow of a gas occurs in a pipeline. Assume uniform flow with the x-direction along the pipe axis and state the simplified continuity equation.

2. Calculate the density gradient in Example 5.1 if (a) a forward difference was used, and (b) if a backward difference was used.

3. The x-component of the velocity vector is measured at three locations 8 mm apart on the centerline of a symmetrical contraction. At points A, B, and C the measurements produce 8.2, 9.4, and 11.1 m/s, respectively. Estimate the y-component of the velocity 2 mm above point B in this steady, plane, incompressible flow.

4. A plane incompressible flow empties radially (no θ component) into a small circular drain. How must the radial component of velocity vary with radius as demanded by continuity?

5. The velocity component $v_\theta = -25(1 + 1/r^2)\sin\theta + 50/r^2$ in a plane incompressible flow. Find $v_r(r, \theta)$ if $v_r(r, 90°) = 0$.

6. Simplify the Navier-Stokes equation for flow in a pipe assuming $v_z = v_z(r)$ and gravity in the z-direction. The streamlines are assumed to be parallel to the pipe wall so that $v_\theta = v_r = 0$.

CHAPTER 6

Dimensional Analysis and Similitude

Many problems of interest in fluid mechanics cannot be solved using the integral and/or differential equations. Wind motions around a football stadium, the air flow around the deflector on a semitruck, the wave motion around a pier or a ship, and air flow around aircraft are all examples of problems which are studied in the laboratory with the use of models. A laboratory study with the use of models is very expensive, however, and to minimize the cost, dimensionless parameters are used. In fact, such parameters are also used in numerical studies for the same reason.

Once an analysis is done on a model in the laboratory and all quantities of interest are measured, it is necessary to predict those same quantities on the prototype, such as the power generated by a large wind machine from measurements on a much smaller model. *Similitude* is the study that allows us to predict the quantities to be expected on a prototype from measurements on a model. This will be done after our study of dimensional analysis that guides the model study.

6.1 Dimensional Analysis

Dimensionless parameters are obtained using a method called *dimensional analysis*. It is based on the idea of *dimensional homogeneity*: all terms in an equation must have the same dimensions. By simply using this idea, we can minimize the number of parameters needed in an experimental or analytical analysis, as will be shown. Any equation can be expressed in terms of dimensionless parameters simply by dividing each term by one of the other terms. For example, consider Bernoulli's equation,

$$\frac{V_2^2}{2} + \frac{p_2}{\rho} + gz_2 = \frac{V_1^2}{2} + \frac{p_1}{\rho} + gz_1 \qquad (6.1)$$

Now, divide both sides by gz_2. The equation can then be written as

$$\frac{V_2^2}{2gz_2} + \frac{p_2}{\gamma z_2} + 1 = \left(\frac{V_1^2}{2gz_1} + \frac{p_1}{\gamma z_1} + 1\right)\frac{z_1}{z_2} \qquad (6.2)$$

Note the dimensionless parameters, V^2/gz and $p/\gamma z$.

Let's use an example to demonstrate the usefulness of dimensional analysis. Suppose the drag force is desired on an object with a spherical front that is shaped as shown in Fig. 6.1. A study could be performed, the drag force measured for a

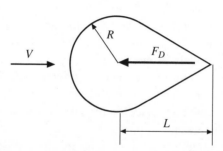

Figure 6.1   Flow around an object.

particular radius R and length L in a fluid with velocity V, viscosity μ, and density ρ. Gravity is expected to not influence the force. This dependence of the drag force on the other variables would be written as

$$F_D = f(R, L, V, \mu, \rho) \tag{6.3}$$

To present the results of an experimental study, the drag force could be plotted as a function of V for various values of the radius R holding all other variables fixed. Then a second plot could show the drag force for various values of L holding all other variables fixed, and so forth. The plots may resemble those of Fig. 6.2. To vary the viscosity holding the density fixed and then the density holding the viscosity fixed, would require a variety of fluids leading to a very complicated study, and perhaps an impossible study.

The actual relationship that would relate the drag force to the other variables could be expressed as a set of dimensionless parameters, much like those of Eq. (6.2), as

$$\frac{F_D}{\rho V^2 R^2} = f\left(\frac{\rho V R}{\mu}, \frac{R}{L}\right) \tag{6.4}$$

(The procedure to do this will be presented next.) The results of a study using the above relationship would be much more organized than the study suggested by the curves of Fig. 6.2. An experimental study would require only several different models, each with different R/L ratios, and only one fluid, either air or water. Varying the velocity of the fluid approaching the model, a rather simple task, could vary the other two dimensionless parameters. A plot of $F_D/(\rho V^2 R^2)$ versus $\rho V R/\mu$ for the several values of R/L would then provide the results of the study.

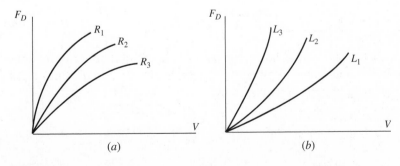

Figure 6.2   Drag force versus velocity: (a) L, μ, ρ fixed; (b) R, μ, ρ fixed.

Before we present the details of forming the dimensionless parameters of Eq. (6.4), let's review the dimensions on quantities of interest in fluid mechanics. Many quantities have obvious dimensions but for some, the dimensions are not so obvious. There are only three basic dimensions, since Newton's second law can be used to relate the basic dimensions. Using F, M, L, and T as the dimensions on force, mass, length, and time, we see that $F = ma$ demands that the dimensions are related by

$$F = M\frac{L}{T^2} \qquad (6.5)$$

We choose to select the *M-L-T* system[1] and use Eq. (6.5) to relate F to M, L, and T. If temperature is needed, as with the flow of a compressible gas, an equation of state, such as

$$p = \rho RT \qquad (6.6)$$

could be expressed dimensionally as

$$[RT] = [p/\rho] = \frac{F}{L^2} \times \frac{L^3}{M} = \frac{ML/T^2}{L^2} \times \frac{L^3}{M} = \frac{L^2}{T^2} \qquad (6.7)$$

where the brackets mean "the dimensions of." Note that the product RT does not introduce additional dimensions.

Table 6.1 has been included to aid in selecting the proper dimensions for quantities of interest. It will simplify the creation of the dimensionless parameters. The dimensions are displayed for the *M-L-T* system only, since that will be what is used in the solution to the problems in this chapter. The same results would be obtained using the *F-L-T* system, should that system be selected.

The *Buckingham π theorem* is used to create the dimensionless parameters, given a functional relationship such as that of Eq. (6.3). Write the primary variable of interest as a general function, such as

$$x_1 = f(x_2, x_2, x_4, \ldots, x_n) \qquad (6.8)$$

where n is the total number of variables. If m is the number of basic dimensions, usually 3, the Buckingham π theorem demands that $(n - m)$ dimensionless groups of variables, the π terms, are related by

$$\pi_1 = f_1(\pi_2, \pi_3, \ldots, \pi_{n-m}) \qquad (6.9)$$

---

[1]The *F-L-T* system could have been used. It's simply our choice to use the *M-L-T* system.

Table 6.1 Symbols and Dimensions of Quantities of
Interest Using the *M-L-T* System

Quantity	Symbol	Dimensions
Length	l	L
Mass	m	M
Time	t	T
Velocity	V	L/T
Acceleration	a	L/T^2
Angular velocity	Ω	T^{-1}
Force	F	F
Gravity	g	L/T^2
Flow rate	Q	L^3/T
Mass flux	$\dot{m}$	M/T
Pressure	p	M/LT^2
Stress	τ	M/LT^2
Density	ρ	M/L^3
Specific weight	γ	M/L^2T^2
Work	W	ML^2/T^2
Viscosity	μ	M/LT
Kinematic viscosity	ν	L^2/T
Power	$\dot{W}$	ML^2/T^3
Heat flux	$\dot{Q}$	ML^2/T^3
Surface tension	σ	M/T^2
Bulk modulus	B	M/LT^2

The π term π_1 is selected to contain the dependent variable [it would be F_D of Eq. (6.3)] and the remaining π terms contain the independent variables. It should be noted that a functional relationship cannot contain a particular dimension in only one variable; for example, in the relationship $v = f(d, t, \rho)$, the density ρ cannot occur since it is the only variable that contains the dimension M, and M would not have the possibility of canceling out to form a dimensionless π term.

The steps that are followed when applying the Buckingham π theorem are:

1. Write the dependent variable as a function of the $(n-1)$ independent variables. This step requires knowledge of the phenomenon being studied. All variables that influence the dependent variable must be included and all variables that do not influence the dependent variable should not be included. In most problems, this relationship will be given.

2. Identify the *m repeating variables* that are combined with the remaining variables to form the π terms. The *m* variables must include all the basic dimensions present in the *n* variables of the functional relationship, but they must not form a dimensionless π term by themselves. Note that an angle is dimensionless, so it is not a candidate to be a repeating variable.

3. Combine each of the $(n-m)$ variables with the repeating variables to form the π terms.

4. Write the π term containing the dependent variable as a function of the remaining π terms.

Step 3 is carried out by either inspection or by an algebraic procedure. The method of inspection will be used in an example. To demonstrate the algebraic procedure, let's form a π term of the variables V, R, ρ, and μ. This is written as

$$\pi = V^a R^b \rho^c \mu^d \tag{6.10}$$

In terms of dimensions, this is

$$M^0 L^0 T^0 = \left(\frac{L}{T}\right)^a L^b \left(\frac{M}{L^3}\right)^c \left(\frac{M}{LT}\right)^d \tag{6.11}$$

Equating exponents on each of the basic dimensions provides the system of equations:

$$
\begin{aligned}
M: & \quad 0 = c + d \\
L: & \quad 0 = a + b - 3c - d \\
T: & \quad 0 = -a - d
\end{aligned}
\tag{6.12}
$$

The solution is

$$c = -d \qquad a = -d \qquad b = -d \tag{6.13}$$

The π term is then written as

$$\pi = \left(\frac{\mu}{VR\rho}\right)^d \tag{6.14}$$

This π term is dimensionless regardless of the value of d. If we desire V to be in the denominator, select $d = 1$; if we desire V to be in the numerator, select $d = -1$. Select $d = -1$ so that

$$\pi = \frac{VR\rho}{\mu} \qquad (6.15)$$

Suppose that only one π term results from an analysis. That π term would then be equal to a constant which could be determined by a single experiment.

Finally, consider a very general functional relationship between a pressure change Δp, a length l, a velocity V, gravity g, viscosity μ, a density ρ, the speed of sound c, the surface tension σ, and an angular velocity Ω. All of these variables may not influence a particular problem, but it is interesting to observe the final relationship of dimensionless terms. Dimensional analysis, using V, l, and ρ as repeating variables provides the relationship

$$\frac{\Delta p}{\rho V^2} = f\left(\frac{V^2}{lg},\ \frac{\rho Vl}{\mu},\ \frac{V}{c},\ \frac{\rho l V^2}{\sigma},\ \frac{\Omega l}{V}\right) \qquad (6.16)$$

Each term that appears in this relationship is an important parameter in certain flow situations. The dimensionless term with its common name is listed as follows:

$$\frac{\Delta p}{\rho V^2} = \text{Eu} \qquad \text{\textbf{Euler number}}$$

$$\frac{V}{\sqrt{lg}} = \text{Fr} \qquad \text{\textbf{Froude number}}$$

$$\frac{\rho Vl}{\mu} = \text{Re} \qquad \text{\textbf{Reynolds number}}$$

$$\frac{V}{c} = \text{M} \qquad \text{\textbf{Mach number}} \qquad (6.17)$$

$$\frac{\rho l V^2}{\sigma} = \text{We} \qquad \text{\textbf{Weber number}}$$

$$\frac{\Omega l}{V} = \text{St} \qquad \text{\textbf{Strouhal number}}$$

Not all of the above numbers would be of interest in a particular flow; it is highly unlikely that both compressibility effects and surface tension would influence the same flow. These are, however, the primary dimensionless parameters in our study of fluid mechanics. The Euler number is of interest in most flows, the Froude number in flows with free surfaces in which gravity is significant (e.g., wave motion), the Reynolds number in flows in which viscous effects are important, the Mach number in compressible flows, the Weber number in flows affected by surface tension (e.g., sprays with droplets), and the Strouhal number in flows in which rotation or a periodic motion plays a role. Each of these numbers, with the exception of the Weber number (surface tension effects are of little engineering importance), will appear in flows studied in subsequent chapters. Note: The Froude number is often defined as V^2/lg; this would not influence the solution to problems.

EXAMPLE 6.1

The pressure drop Δp over a length L of pipe is assumed to depend on the average velocity V, the pipe's diameter D, the average height e of the roughness elements of the pipe wall, the fluid density ρ, and the fluid viscosity μ. Write a relationship between the pressure drop and the other variables.

Solution

First, select the repeating variables. Do not select Δp since that is the dependent variable. Select only one D, L, and e since they all have the dimensions of length. Select the variables that are thought[2] to most influence the pressure drop: V, D, and ρ. Now, list the dimensions on each variable (refer to Table 6.1)

$$[\Delta p] = \frac{M}{LT^2} \quad [L] = L \quad [V] = \frac{L}{T} \quad [D] = L$$

$$[e] = L \quad [\rho] = \frac{M}{L^3} \quad [\mu] = \frac{M}{LT}$$

First, combine Δp, V, D, and μ into a π term. Since only Δp and ρ have M as a dimension, they must occur as a ratio $\Delta p/\rho$. That places T in the denominator so that V must be in the numerator so the T's cancel out. Finally, check out the L's: there is L^2 in the numerator so D^2 must be in the denominator providing

$$\pi_1 = \frac{\Delta p}{\rho V^2 D^2}$$

---

[2]This is often debatable. Either D or L could be selected, whichever is considered to be most influential.

The second π term is found by combining L with the three repeating variables V, D, and ρ. Since both L and D have the dimension of length, the second π term is

$$\pi_2 = \frac{L}{D}$$

The third π term results from combining e with the repeating variables. It has the dimension of length so the third π term is

$$\pi_3 = \frac{e}{D}$$

The last π term is found by combining μ with V, D, and ρ. Both μ and ρ contain the dimension M demanding that they form the ratio ρ/μ. This puts T in the numerator demanding that V goes in the numerator. This puts L in the denominator so that D must appear in the numerator. The last π term is then

$$\pi_4 = \frac{\rho V D}{\mu}$$

The final expression relates the π terms as $\pi_1 = f(\pi_2, \pi_3, \pi_4)$ or, using the variables,

$$\frac{\Delta p}{\rho V^2 D^2} = f\left(\frac{L}{D}, \frac{e}{D}, \frac{\rho V D}{\mu} \right)$$

If L had been chosen as a repeating variable, it would simply change places with D since it has the same dimension.

EXAMPLE 6.2
The speed V of a weight when it hits the floor is assumed to depend on gravity g, the height h from which it was dropped, and the density ρ of the weight. Use dimensional analysis and write a relationship between the variables.

Solution
The dimensions of each variable are listed as

$$[V] = \frac{L}{T} \qquad [g] = \frac{L}{T^2} \qquad [h] = L \qquad [\rho] = \frac{M}{L^3}$$

Since M occurs in only one variable, that variable ρ cannot be included in the relationship. The remaining three terms are combined to form a single π term; it is formed by observing that T occurs in only two of the variables, thus V^2 is in the numerator and g is in the denominator. The length dimension is then canceled by placing h in the denominator. The single π term is

$$\pi_1 = \frac{V^2}{gh}$$

Since this π term depends on all other π terms and there are none, it must be at most a constant. Hence, we conclude that

$$V = C\sqrt{gh}$$

A simple experiment would show that $C = \sqrt{2}$. We see that dimensional analysis rules out the possibility that the speed of free fall, neglecting viscous effects (e.g., drag), depends on the density of the material (or the weight).

6.2 Similitude

After the dimensionless parameters have been identified and a study on a model has been accomplished in a laboratory, *similitude* allows us to predict the behavior of a prototype from the measurements made on the model. The measurements on the model of a ship in a towing basin or on the model of an aircraft in a wind tunnel are used to predict the performance of the ship or the aircraft.

The application of similitude is based on three types of similarity. First, a model must look like the prototype, that is, the length ratio must be constant between corresponding points on the model and prototype. For example, if the ratio of the lengths of the model and prototype is λ, then every other length ratio is also λ. Hence, the area ratio would be λ^2 and the volume ratio λ^3. This is *geometric similarity*.

The second is *dynamic similarity*: all force ratios acting on corresponding mass elements in the model flow and the prototype flow are the same. This results by equating the appropriate dimensionless numbers of Eq. (6.17). If viscous effects are important, the Reynolds numbers are equated; if compressibility is significant, Mach numbers are equated; if gravity influences the flows, Froude numbers are equated; if an angular velocity influences the flow, the Strouhal numbers are equated; and, if surface tension affects the flow, the Weber numbers are equated. All of these numbers can be shown to be ratios of forces so equating the numbers in a particular flow is equivalent to equating the force ratios in that flow.

The third type of similarity is *kinematic similarity*: the velocity ratio is the same between corresponding points in the flow around the model and the prototype.

Assuming complete similarity between model and prototype, quantities of interest can now be predicted. For example, if a drag force is measured on flow around a model in which viscous effects play an important role, the ratio of the forces [see Eq. (6.18)] would be

$$\frac{(F_D)_m}{(F_D)_p} = \frac{\rho_m V_m^2 l_m^2}{\rho_p V_p^2 l_p^2} \tag{6.18}$$

The velocity ratio would be found by equating the Reynolds numbers:

$$\text{Re}_m = \text{Re}_p \qquad \frac{\rho_m V_m l_m}{\mu_m} = \frac{\rho_p V_p l_p}{\mu_p} \tag{6.19}$$

If the length ratio, the *scale*, is given and the same fluid is used in model and prototype ($\rho_m = \rho_p$ and $\mu_m = \mu_p$), the force acting on the prototype can be found. It would be

$$(F_D)_p = (F_D)_m \left(\frac{V_p}{V_m}\right)^2 \left(\frac{l_p}{l_m}\right)^2 = (F_D)_m \left(\frac{l_m}{l_p}\right)^2 \left(\frac{l_p}{l_m}\right)^2 = (F_D)_m \tag{6.20}$$

showing that, if the Reynolds number governs the model study and the same fluid is used in the model and prototype, the force on the model is the same as the force of the prototype. Note that the velocity in the model study is the velocity in the prototype multiplied by the length ratio so that the model velocity could be quite large.

If the Froude number governed the study, we would have

$$\text{Fr}_m = \text{Fr}_p \qquad \frac{V_m^2}{l_m g_m} = \frac{V_p^2}{l_p g_p} \tag{6.21}$$

The drag force on the prototype, with $g_m = g_p$, would then be

$$(F_D)_p = (F_D)_m \left(\frac{V_p}{V_m}\right)^2 \left(\frac{l_p}{l_m}\right)^2 = (F_D)_m \left(\frac{l_p}{l_m}\right) \left(\frac{l_p}{l_m}\right)^2 = (F_D)_m \left(\frac{l_p}{l_m}\right)^3 \tag{6.22}$$

This is the situation for the model study of a ship. The Reynolds number is not used even though the viscous drag force acting on the ship cannot be neglected. We cannot satisfy both the Reynolds number and the Froude number in a study if the same fluid is used for the model study as exists in the prototype flow; the model study of a ship always uses water as the fluid. To account for the viscous drag, the results of the model study based on the Froude number are adapted using industrial modifiers not included in an introductory fluids course.

EXAMPLE 6.3

A clever design of the front of a ship is to be tested in a water basin. A drag of 12.2 N is measured on the 1:20 scale model when towed at a speed of 3.6 m/s. Determine the corresponding speed of the prototype ship and the drag to be expected.

Solution

The Froude number guides the model study of a ship since gravity effects (wave motions) are more significant than the viscous effects. Consequently,

$$\text{Fr}_p = \text{Fr}_m \quad \text{or} \quad \frac{V_p^2}{l_p g_p} = \frac{V_m^2}{l_m g_m}$$

Since gravity does not vary significantly on the earth, there results

$$V_p = V_m \sqrt{\frac{l_p}{l_m}} = 3.6 \times \sqrt{20} = 16.1 \text{ m/s}$$

To find the drag on the prototype, the drag ratio is equated to the gravity force ratio (the inertial force ratio could be used but not the viscous force ratio since viscous forces have been ignored):

$$\frac{(F_D)_p}{(F_D)_m} = \frac{\rho_p V_p^2 l_p^2}{\rho_m V_m^2 l_m^2} \quad \therefore (F_D)_p = (F_D)_m \frac{V_p^2 l_p^2}{V_m^2 l_m^2} = 12.2 \times \frac{16.1^2}{3.6^2} \times 20^2 = 97\,600 \text{ N}$$

where we used $\rho_m = \rho_p$ since salt water and fresh water have nearly the same density. The above results would be modified based on established factors to account for the viscous drag on the ship.

EXAMPLE 6.4

A large pump delivering 1.2 m³/s of water with a pressure rise of 400 kPa is needed for a particular hydroelectric power plant. A proposed design change is tested on a smaller 1:4 scale pump. Estimate the flow rate and pressure rise that would be expected in the model study. If the power needed to operate the model pump is measured to be 8000 kW, what power would be expected to operate the prototype pump?

Solution

For this internal flow problem, Reynolds number would be equated:

$$\text{Re}_p = \text{Re}_m \qquad \text{or} \qquad \frac{V_p d_p}{\nu_p} = \frac{V_m d_m}{\nu_m} \qquad \text{or} \qquad \frac{V_p}{V_m} = \frac{d_m}{d_p}$$

assuming $\nu_p \cong \nu_m$ for the water in the model and prototype. The ratio of flow rates is

$$\frac{Q_p}{Q_m} = \frac{A_p V_p}{A_m V_m} = \frac{l_p^2 V_p}{l_m^2 V_m} = 4^2 \times \frac{1}{4} = 4 \qquad \therefore Q_m = \frac{Q_p}{4} = \frac{1.2}{4} = 0.3 \text{ m}^3\text{/s}$$

The power ratio is found by using power as force times velocity; this provides

$$\frac{\dot{W}_p}{\dot{W}_m} = \frac{\rho_p V_p^2 l_p^2}{\rho_m V_m^2 l_m^2} \times \frac{V_p}{V_m} \qquad \therefore \dot{W}_p = \dot{W}_m \left(\frac{d_m}{d_p}\right)^2 \left(\frac{d_p}{d_m}\right)^2 \times \frac{d_m}{d_p} = \frac{8000}{4} = 2000 \text{ kW}$$

This is an unexpected result. When using the Reynolds number to guide a model study, the power measured on the model exceeds the power needed to operate the prototype since the pressures are so much larger on the model. Note that in this example the Euler number would be used to provide the model pressure rise as

$$\frac{\Delta p_p}{\Delta p_m} = \frac{\rho_p V_p^2}{\rho_m V_m^2} \qquad \therefore \Delta p_m = \Delta p_p \times \left(\frac{d_p}{d_m}\right)^2 = 400 \times 4^2 = 6400 \text{ kPa}$$

For this reason and the observation that the velocity is much larger on the model, model studies are not common for situations (e.g., flow around an automobile) in which the Reynolds number is the guiding parameter.

EXAMPLE 6.5

The pressure rise from free stream to a certain location on the surface of the model of a rocket is measured to be 22 kPa at an air speed of 1200 km/h. The wind tunnel is maintained at 90 kPa absolute and 15°C. What would be the speed and pressure rise on a rocket prototype at an elevation of 15 km?

Solution

The Mach number governs the model study. Thus,

$$M_m = M_p \qquad \frac{V_m}{c_m} = \frac{V_p}{c_p} \qquad \frac{V_m}{\sqrt{kRT_m}} = \frac{V_p}{\sqrt{kRT_p}}$$

Using the temperature from Table C.3, the velocity is

$$V_p = V_m \sqrt{\frac{T_p}{T_m}} = 1200 \sqrt{\frac{216.7}{288}} = 1041 \text{ km/h}$$

A pressure force is $\Delta p A \approx \Delta p l^2$ so that the ratio to the inertial force of Eq. (6.18) is the Euler number, $\Delta p / \rho V^2$. Equating the Euler numbers gives the pressure rise as

$$\Delta p_p = \Delta p_m \frac{\rho_p V_p^2}{\rho_m V_m^2} = 22 \frac{p_p T_m V_p^2}{p_m T_p V_m^2} = 22 \frac{12.3 \times 288 \times 1041^2}{90 \times 216.7 \times 1200^2} = 3.01 \text{ kPa}$$

Quiz No. 1

1. If the *F-L-T* system is used, the dimensions on density are:

    (A) FT/L^2

    (B) FT^2/L^2

    (C) FT/L^4

    (D) FT^2/L^4

2. Combine the angular velocity ω, viscosity μ, diameter d, and density ρ into a single dimensionless group, a π term.

    (A) $\omega\rho b/\mu$

    (B) $\omega\rho b^4/\mu$

    (C) $\omega\rho b^2/\mu$

    (D) $\omega^2\rho b/\mu$

3. What variable could not influence the velocity if it is proposed that the velocity depends on a diameter, a length, gravity, rotational speed, and viscosity?

   (A) Gravity

   (B) Rotational speed

   (C) Viscosity

   (D) Diameter

4. It is proposed that the velocity V issuing from a hole in the side of an open tank depends on the density ρ of the fluid, the distance H from the surface, and gravity g. What expression relates the variables?

   (A) $V = C\sqrt{gH}$

   (B) $V = \sqrt{\rho g H}$

   (C) $V = gH/\rho$

   (D) $V = C\sqrt{\rho g H}$

5. Add viscosity to the variables in Prob. 4. Rework the problem.

   (A) $V/\sqrt{gH} = f\left(\rho\sqrt{gH^3}/\mu\right)$

   (B) $V/\sqrt{gH} = f\left(\rho\sqrt{gH^3}/\mu\right)$

   (C) $V/\sqrt{\rho gH} = f\left(\rho\sqrt{gH^3}/\mu\right)$

   (D) $V/\sqrt{\rho gH} = f\left(\mu\sqrt{gH^3}/\rho\right)$

6. The lift F_L on an airfoil is related to its velocity V, its length L, its chord length c, its angle of attack α, and the density ρ of the air. Viscous effects are assumed negligible. Relate the lift to the other variables.

   (A) $F_L/\rho Vc^2 = f(c/L, \alpha)$

   (B) $F_L/\rho Vc = f(c/L, \alpha c/L)$

   (C) $F_L/\rho V^2 c = f(c/L, \alpha/L)$

   (D) $F_L/\rho V^2 c^2 = f(c/L, \alpha)$

7. A new design is proposed for an automobile. It is suggested that a 1:5 scale model study be done to access the proposed design for a speed of 90 km/h. What speed should be selected for the model study, based on the appropriate parameter?

   (A) 500 km/h

   (B) 450 km/h

   (C) 325 km/h

   (D) 270 km/h

8. A proposed pier design is studied in a water channel to simulate forces due to hurricanes. Using a 1:10 scale model, what velocity should be selected in the model study to simulate a water speed of 12 m/s?

   (A) 3.79 m/s

   (B) 4.28 m/s

   (C) 5.91 m/s

   (D) 6.70 m/s

9. The force on a weir is to be predicted by studying the flow of water over a 1:10 scale model. If 1.8 m³/s is expected over the weir, what flow rate should be used in the model study?

   (A) 362 m³/s

   (B) 489 m³/s

   (C) 569 m³/s

   (D) 674 m³/s

10. What force should be expected on the weir of Prob. 9 if 20 N is measured on the model?

   (A) 15 kN

   (B) 20 kN

   (C) 25 kN

   (D) 30 kN

Quiz No. 2

1. If the *F-L-T* system is used, what are the dimensions on viscosity?

2. Combine power $\dot{W}$, diameter *d*, velocity *V*, and pressure rise Δp into a single dimensionless group, a π term.

3. The speed *V* of a weight when it hits the floor is assumed to depend on gravity *g*, the height *h* from where it was dropped, and the density ρ of the weight. Use dimensional analysis and write a relationship between the variables.

4. An object falls freely in a viscous fluid. Relate the terminal velocity *V* to its width *w*, its length *l*, gravity *g*, and the fluid density ρ and viscosity μ. Select *w*, ρ, and μ as repeating variables.

5. The pressure drop Δp over a horizontal section of pipe of diameter d depends on the average velocity V, the viscosity μ, the fluid density ρ, the average height ε of the surface roughness elements, and the length L of the pipe section. Write an expression that relates the pressure drop to the other variables. Select d, V, and ρ as repeating variables.

6. The drag force F_D on a sphere depends on the sphere's diameter d and velocity V, the fluid's viscosity μ and density ρ, and gravity g. Write an expression for the drag force.

7. A model of a golf ball is to be studied to determine the effects of the dimples. A sphere 10 times larger than an actual golf ball is used in the wind tunnel study. What speed should be selected for the model to simulate a prototype speed of 50 m/s?

8. A towing force of 15 N is measured on a 1:40 scale model of a ship in a water channel. What velocity should be used to simulate a prototype speed of 10 m/s? What would be the predicted force on the ship at that speed?

9. A 1:20 scale model of an aircraft is studied in a 20°C supersonic wind tunnel at sea level. If a lift of 200 N at a speed of 250 m/s is measured in the wind tunnel, what velocity and lift does that simulate for the prototype? Assume the prototype is at 4000 m.

CHAPTER 7

Internal Flows

The material in this chapter is focused on the influence of viscosity on the flows internal to boundaries, such as flow in a pipe or between rotating cylinders. The next chapter will focus on flows that are external to a boundary, such as an airfoil. The parameter that is of primary interest in an internal flow is the Reynolds number, Re = VL/v, where L is the primary characteristic length (e.g., the diameter of a pipe) in the problem of interest and V is usually the average velocity in a flow. We will consider internal flows in pipes, between parallel plates and rotating cylinders, and in open channels. If the Reynolds number is relatively low, the flow is laminar (see Sec. 3.2); if it is relatively high, it is turbulent. For pipe flows, the flow is assumed to be laminar if Re < 2000; for flow between wide parallel plates, it is laminar if Re < 1500; for flow between rotating concentric cylinders, it is laminar and flows in a circular motion below Re < 1700; and in the open channels of interest, it is assumed to be turbulent.

7.1 Entrance Flow

The flows mentioned above refer to *developed flows*, flows in which the velocity profiles do not change in the streamwise direction. In the region near a geometry

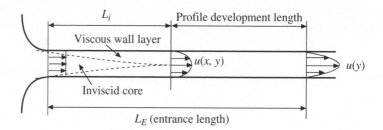

Figure 7.1　The laminar-flow entrance region in a pipe or between parallel plates.

change, such as an elbow, a valve, or near an entrance, the velocity profile changes in the flow direction. Let's consider the changes in the entrance region for a laminar flow in a pipe or between parallel plates. The *entrance length* L_E is shown in Fig. 7.1. The velocity profile very near the entrance is essentially uniform, the *viscous wall layer* grows until it permeates the entire cross section over the *inviscid core length* L_i; the profile continues to develop into a developed flow at the end of the *profile development region*.

For a laminar flow in a pipe, with a uniform velocity profile at the entrance,

$$\frac{L_E}{D} = 0.065\,\mathrm{Re} \qquad \mathrm{Re} = \frac{VD}{v} \tag{7.1}$$

where V is the average velocity and D is the diameter. The inviscid core is about half of the entrance length.[1] It should be mentioned that laminar flows in pipes have been observed at Reynolds numbers as high as 40 000 in extremely controlled flows in smooth pipes in a building free of vibrations; for a conventional pipe with a rough wall we use 2000 as the limit[2] for a laminar flow.

For flow between wide parallel plates, with a uniform profile at the entrance,

$$\frac{L_E}{h} = 0.04\,\mathrm{Re} \qquad \mathrm{Re} = \frac{Vh}{v} \tag{7.2}$$

where h is the distance between the plates and V is the average velocity. A laminar flow cannot exist for Re > 7700; a value of 1500 is used as the limit for flow in a conventional channel.

The entrance region for a developed turbulent flow is displayed in Fig. 7.2. The velocity profile is developed at the length L_d but the characteristics of the turbulence

---

[1]Some texts may suggest that the inviscid core makes up the entire entrance region, an assumption that is not true.

[2]Some texts suggest 2300. This difference does not significantly influence design considerations.

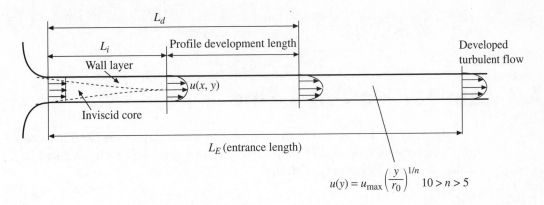

$$u(y) = u_{max}\left(\frac{y}{r_0}\right)^{1/n} \quad 10 > n > 5$$

Figure 7.2   The turbulent-flow entrance region in a pipe.

in the flow require the additional length. For large Reynolds numbers exceeding 10^5 in a pipe, we use

$$\frac{L_i}{D} \cong 10 \qquad \frac{L_d}{D} \cong 40 \qquad \frac{L_E}{D} \cong 120 \tag{7.3}$$

For a flow with Re = 4000 the lengths are possibly five times those listed in Eq. (7.3), due to the initial laminar development followed by the development of turbulence. (Detailed results have not been reported for flows in which Re < 10^5.)

The pressure variation is shown in Fig. 7.3. The initial transition to turbulence from the wall of the pipe is noted in the figure. The pressure variation for the laminar

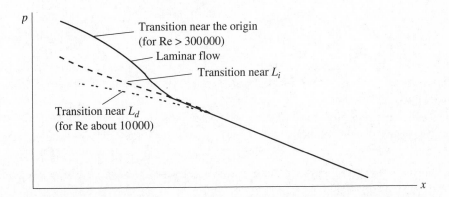

Figure 7.3   Pressure variation in a pipe for both laminar and turbulent flows.

flow is higher in the entrance region than in the fully-developed region due to the larger wall shear and the increasing momentum flux.

7.2 Laminar Flow in a Pipe

Steady, developed laminar flow in a pipe will be derived applying Newton's second law to the element of Fig. 7.4 or using the appropriate Navier-Stokes equation of Chap. 5. Either derivation can be used.

THE ELEMENTAL APPROACH

The element of fluid shown in Fig.7.4 can be considered a control volume into and from which the fluid flows or it can be considered a mass of fluid at a particular moment. Considering it to be an instantaneous mass of fluid that is not accelerating in this steady, developed flow, Newton's second law takes the form

$$\Sigma F_x = 0 \quad \text{or} \quad p\pi r^2 - (p+dp)\pi r^2 - \tau 2\pi r\, dx + \gamma\, \pi r^2\, dx \sin\theta = 0 \quad (7.4)$$

where τ is the shear on the wall of the element and γ is the specific weight of the fluid. This simplifies to

$$\tau = -\frac{r}{2}\frac{d}{dx}(p+\gamma h) \quad (7.5)$$

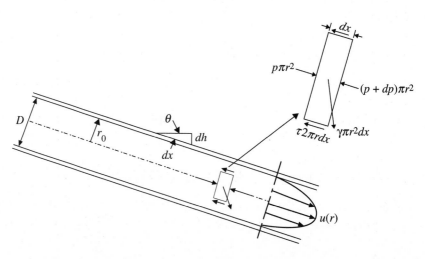

Figure 7.4   Steady, developed flow in a pipe.

using $dh = -\sin\theta dx$ with h measured in the vertical direction. Note that this equation can be applied to either a laminar or a turbulent flow. For a laminar flow, the shear stress τ is related to the velocity gradient[3] by Eq. (1.13):

$$-\mu\frac{du}{dr} = -\frac{r}{2}\frac{d}{dx}(p+\gamma h) \tag{7.6}$$

Because we assume a developed flow (no change of the velocity profile in the flow direction), the left-hand side of the equation is a function of r only, so $d/dx(p+\gamma h)$ must be at most a constant (it cannot depend on r since there is no radial acceleration; we assume the pipe is relatively small, so there is no variation of pressure with r). Hence, we can write

$$\int du = \int \frac{r}{2\mu}\frac{d}{dx}(p+\gamma h)dr \tag{7.7}$$

This is integrated to provide the velocity profile:

$$u(r) = \frac{r^2}{4\mu}\frac{d}{dx}(p+\gamma h) + C \tag{7.8}$$

where the constant of integration C can be evaluated using $u(r_0) = 0$ so that

$$u(r) = \frac{(r^2 - r_0^2)}{4\mu}\frac{d}{dx}(p+\gamma h) \tag{7.9}$$

For a horizontal pipe for which $dh/dx = 0$ the velocity profile becomes

$$u(r) = \frac{1}{4\mu}\frac{dp}{dx}(r^2 - r_0^2) \tag{7.10}$$

The above velocity profile is a parabolic profile; the flow is sometimes referred to as a *Poiseuille flow*.

The same result can be obtained by solving the appropriate Navier-Stokes equation. If that is not of interest, skip the next part.

APPLYING THE NAVIER-STOKES EQUATIONS

The z-component differential momentum equation using cylindrical coordinates from Table 5.1 is applied to a steady, developed flow in a circular pipe. For the present

---

[3]The minus sign is required since the stress is a positive quantity and du/dr is negative.

situation we wish to refer to the coordinate in the flow direction as x and the velocity component in the x-direction as $u(x)$; so, let's replace the z with x and the v_z with u. Then the differential equation takes the form

$$\rho\left(\cancel{v_r\frac{\partial u}{\partial r}} + \cancel{\frac{v_\theta}{r}\frac{\partial u}{\partial \theta}} + u\cancel{\frac{\partial u}{\partial x}} + \cancel{\frac{\partial u}{\partial t}}\right) = -\frac{\partial p}{\partial x} + \rho g_x + \mu\left(\frac{\partial^2 u}{\partial r^2} + \frac{1}{r}\frac{\partial u}{\partial r} + \frac{1}{r^2}\cancel{\frac{\partial^2 u}{\partial \theta^2}} + \cancel{\frac{\partial^2 u}{\partial x^2}}\right)$$

| no radial velocity | no swirl | developed flow | steady flow | | symmetric flow | developed flow |

Observe that the left-hand side of the equation is zero, i.e., the fluid particles are not accelerating. Using $\rho g_x = \gamma\sin\theta = -\gamma\,dh/dx$, the above equation simplifies to

$$\frac{1}{\mu}\frac{\partial}{\partial x}(p+\gamma h) = \frac{1}{r}\frac{\partial}{\partial r}\left(r\frac{\partial u}{\partial r}\right) \qquad (7.11)$$

where we observed that we could write

$$\frac{\partial^2 u}{\partial r^2} + \frac{1}{r}\frac{\partial u}{\partial r} = \frac{1}{r}\frac{\partial}{\partial r}\left(r\frac{\partial u}{\partial r}\right) \qquad (7.12)$$

Now, we see that the left-hand side of Eq. (7.11) is at most a function of x, and the right-hand side is a function of r. This means that each side is at most a constant, say λ, since x and r can be varied independently of each other. So, we replace the partial derivatives with ordinary derivatives and write Eq. (7.11) as

$$\lambda = \frac{1}{r}\frac{d}{dr}\left(r\frac{du}{dr}\right) \qquad \text{or} \qquad d\left(r\frac{du}{dr}\right) = \lambda r\,dr \qquad (7.13)$$

This is integrated to provide

$$r\frac{du}{dr} = \lambda\frac{r^2}{2} + A \qquad (7.14)$$

Multiply by dr/r and integrate again. We have

$$u(r) = \lambda\frac{r^2}{4} + A\ln r + B \qquad (7.15)$$

Refer to Fig. 7.4: the two boundary conditions are, u is finite at $r = 0$, and $u = 0$ at $r = r_0$. Thus, $A = 0$ and $B = -\lambda r_0^2/4$.   Since λ is the left-hand side of Eq. (7.11), we can write Eq. (7.15) as

$$u(r) = \frac{1}{4\mu}\frac{d}{dx}(p + \gamma h)(r^2 - r_0^2) \tag{7.16}$$

This is the parabolic velocity distribution of a developed laminar flow in a pipe, sometimes called a *Poiseuille flow*. For a horizontal pipe, $dh/dx = 0$ and

$$u(r) = \frac{1}{4\mu}\frac{dp}{dx}(r^2 - r_0^2) \tag{7.17}$$

QUANTITIES OF INTEREST

The first quantity of interest in the flow in a pipe is the average velocity V. If we express the constant-pressure gradient as $dp/dx = \Delta p/L$ where Δp is the pressure drop (a positive number) over the length L, there results

$$V = \frac{1}{A}\int u(r)\,2\pi r\,dr$$

$$= -\frac{2\pi}{\pi r_0^2}\frac{\Delta p}{4\mu L}\int_0^{r_0}(r^2 - r_0^2)\,r\,dr = \frac{r_0^2\Delta p}{8\mu L} \tag{7.18}$$

The maximum velocity occurs at $r = 0$ and is

$$u_{\max} = \frac{r_0^2\Delta p}{4\mu L} = 2V \tag{7.19}$$

Rewriting Eq. (7.19), the pressure drop is

$$\Delta p = \frac{8\mu L V}{r_0^2} \tag{7.20}$$

The shear stress at the wall can be found by considering a control volume of length L in the pipe. For a horizontal pipe the pressure force balances the shear force so that the control volume yields

$$\pi r_0^2 \times \Delta p = 2\pi r_0 L \times \tau_0 \qquad \therefore \tau_0 = \frac{r_0\Delta p}{2L} \tag{7.21}$$

Sometimes a dimensionless wall shear called the *friction factor f* is used. It is defined to be

$$f = \frac{\tau_0}{\frac{1}{8}\rho V^2} \tag{7.22}$$

We also refer to a *head loss* h_L defined as $\Delta p/\gamma$. By combining the above equations, it can be expressed as the *Darcy-Weisbach equation*:

$$h_L = \frac{\Delta p}{\gamma} = f\frac{L}{D}\frac{V^2}{2g} \tag{7.23}$$

It is valid for both a laminar and a turbulent flow in a pipe. In terms of the Reynolds number, the friction factor for a laminar flow is [combining Eqs. (7.20) and (7.23)]

$$f = \frac{64}{\text{Re}} \tag{7.24}$$

where Re = VD/ν. If this is substituted into Eq. (7.23), we see that the head loss is directly proportional to the average velocity in a laminar flow, which is also true of a laminar flow in a conduit of any cross section.

EXAMPLE 7.1

The pressure drop over a 30-m length of 1-cm-diameter horizontal pipe transporting water at 20°C is measured to be 2 kPa. A laminar flow is assumed. Determine (a) the maximum velocity in the pipe, (b) the Reynolds number, (c) the wall shear stress, and (d) the friction factor.

Solution

(a) The maximum velocity is found to be

$$u_{max} = \frac{r_0^2 \Delta p}{4\mu L} = \frac{0.005^2 \times 2000}{4\times 10^{-3} \times 30} = 0.4167 \text{ m/s}$$

Note: The pressure must be in pascals in order for the units to check. It is wise to make sure the units check when equations are used for the first time. The above units are checked as follows:

$$\frac{m^2 \times N/m^2}{(N\cdot s/m^2)\times m} = m/s$$

(b) The Reynolds number, a dimensionless quantity, is (use $V = u_{max}/2$)

$$\text{Re} = \frac{VD}{\nu} = \frac{(0.4167/2) \times 0.01}{10^{-6}} = 2080$$

This exceeds 2000 but a laminar flow can exist at higher Reynolds numbers if a smooth pipe is used and care is taken to provide a flow free of disturbances. Note how low the velocity is in this relatively small pipe. Laminar flows do not exist in most engineering applications unless the fluid is extremely viscous or the dimensions are quite small.

(c) The wall shear stress due to the viscous effects is found to be

$$\tau_0 = \frac{r_0 \Delta p}{2L} = \frac{0.005 \times 2000}{2 \times 30} = 0.1667 \text{ Pa}$$

If we had used the pressure in kPa, the stress would have had units of kPa.

(d) Finally, the friction factor, a dimensionless quantity, is

$$f = \frac{\tau_0}{\rho V^2/2} = \frac{0.1667}{0.5 \times 1000 \times (0.4167/2)^2} = 0.00768$$

7.3 Laminar Flow Between Parallel Plates

Steady, developed laminar flow between parallel plates (the top plate moving with velocity U) will be derived applying Newton's second law to the element of Fig. 7.5 or using the appropriate Navier-Stokes equation of Chap. 5. Either derivation can be used.

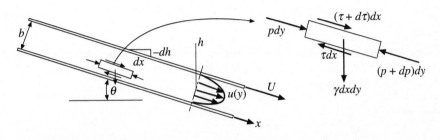

Figure 7.5   Steady, developed flow between parallel plates.

THE ELEMENTAL APPROACH

The element of fluid shown in Fig. 7.5 can be considered a control volume into and from which the fluid flows or it can be considered a mass of fluid at a particular moment. Considering it to be an instantaneous mass of fluid that is not accelerating in this steady, developed flow, Newton's second law takes the form

$$\Sigma F_x = 0 \qquad \text{or} \qquad p\,dy - (p+dp)\,dy + \tau\,dx - (\tau+d\tau)\,dx + \gamma\,dx\,dy\sin\theta = 0 \quad (7.25)$$

where τ is the shear on the wall of the element and γ is the specific weight of the fluid. We have assumed a unit length into the paper (in the z-direction). To simplify, divide by $dxdy$ and use $dh = -\sin\theta\,dx$ with h measured in the vertical direction:

$$\frac{d\tau}{dy} = \frac{d}{dx}(p+\gamma h) \qquad (7.26)$$

For this laminar flow the shear stress is related to the velocity gradient by $\tau = \mu du/dy$ so that Eq. (7.26) becomes

$$\mu\frac{d^2u}{dy^2} = \frac{d}{dx}(p+\gamma h) \qquad (7.27)$$

The left-hand side of the equation is a function of y only for this developed flow (we assume a wide channel with an aspect ratio in excess of 8) and the right-hand side is a function of x only. So, we can integrate twice on y to obtain

$$u(y) = \frac{1}{2\mu}\frac{d(p+\gamma h)}{dx}y^2 + Ay + B \qquad (7.28)$$

Using the boundary conditions $u(0) = 0$ and $u(b) = U$, the constants of integration are evaluated and a parabolic profile results:

$$u(y) = \frac{1}{2\mu}\frac{d(p+\gamma h)}{dx}(y^2 - by) + \frac{U}{b}y \qquad (7.29)$$

If the plates are horizontal and $U = 0$, the velocity profile simplifies to

$$u(y) = \frac{\Delta p}{2\mu L}(by - y^2) \qquad (7.30)$$

where we have let $d(p + \gamma h)/dx = -\Delta p/L$ for the horizontal plates, where Δp is the pressure drop, a positive quantity.

If the flow is due only to the top horizontal plate moving with velocity U, with zero pressure gradient, it is a *Couette flow* so that $u(y) = Uy/b$. If both plates are stationary and the flow is due only to a pressure gradient, it is a Poiseuille flow.

The same result can be obtained by solving the appropriate Navier-Stokes equation. If that is not of interest, skip the next part.

APPLYING THE NAVIER-STOKES EQUATIONS

The x-component differential momentum equation in rectangular coordinates [see Eq. (5.18)] is selected for this steady, developed flow with streamlines parallel to the walls in a wide channel (at least an 8 to 1 aspect ratio):

$$\rho\left(\cancel{\frac{\partial u}{\partial t}} + u\cancel{\frac{\partial u}{\partial x}} + \cancel{v}\frac{\partial u}{\partial y} + \cancel{w}\frac{\partial u}{\partial z} \right) = -\frac{\partial p}{\partial x} + \gamma\sin\theta + \mu\left(\cancel{\frac{\partial^2 u}{\partial x^2}} + \frac{\partial^2 u}{\partial y^2} + \cancel{\frac{\partial^2 u}{\partial z^2}} \right) \quad (7.31)$$

$$\underset{\substack{\text{steady} \quad \text{developed} \quad \text{streamlines parallel} \\ \text{to wall}}}{} \qquad\qquad \underset{\text{developed} \quad \text{wide channel}}{}$$

where the channel makes an angle of θ with the horizontal. Using $dh = -dx\sin\theta$ this partial differential equation simplifies to

$$\frac{d^2 u}{dy^2} = \frac{1}{\mu}\frac{d}{dx}(p + \gamma h) \qquad (7.32)$$

where the partial derivatives have been replaced by ordinary derivatives since u depends on y only and p is a function of x only.

Because the left-hand side of Eq. (7.32) is a function of y and the right-hand side is a function of x, both of which can be varied independent of each other, the two sides can be at most a constant, say λ, so that

$$\frac{d^2 u}{dy^2} = \lambda \qquad (7.33)$$

Integrating twice provides

$$u(y) = \frac{1}{2}\lambda y^2 + Ay + B \qquad (7.34)$$

Refer to Fig. 7.5: the boundary conditions are $u(0) = 0$ and $u(b) = U$ providing

$$A = \frac{U}{b} - \lambda\frac{b}{2} \qquad B = 0 \qquad (7.35)$$

The velocity profile is thus

$$u(y) = \frac{d(p+\gamma h)/dx}{2\mu}(y^2 - by) + \frac{U}{b}y \tag{7.36}$$

where λ has been used as the right-hand side of Eq. (7.32).

In a horizontal channel we can write $d(p + \lambda h)/dx = -\Delta p/L$. If $U = 0$ the velocity profile is

$$u(y) = \frac{\Delta p}{2\mu L}(by - y^2) \tag{7.37}$$

This is a Poiseuille flow. If the pressure gradient is zero and the motion of the top plate causes the flow, it is called a *Couette flow* with $u(y) = Uy/b$.

QUANTITIES OF INTEREST

Let us consider several quantities of interest for the case of two fixed plates with $U = 0$. The first quantity of interest in the flow is the average velocity V. The average velocity is, assuming unit width of the plates

$$V = \frac{1}{b}\int u(y)\,dy$$

$$= \frac{\Delta p}{2b\mu L}\int_0^b (by - y^2)\,dr = \frac{\Delta p}{2b\mu L}\left(b\frac{b^2}{2} - \frac{b^3}{3}\right) = \frac{b^2 \Delta p}{12\mu L} \tag{7.38}$$

The maximum velocity occurs at $y = b/2$ and is

$$u_{max} = \frac{\Delta p}{2\mu L}\left(\frac{b^2}{2} - \frac{b^2}{4}\right) = \frac{b^2 \Delta p}{8\mu L} = \frac{2}{3}V \tag{7.39}$$

The pressure drop, rewriting Eq. (7.38), for this horizontal[4] channel, is

$$\Delta p = \frac{12\mu LV}{b^2} \tag{7.40}$$

---

[4]For a sloped channel simply replace p with $(p + \gamma h)$.

The shear stress at either wall can be found by considering a free body of length L in the channel. For a horizontal channel the pressure force balances the shear force:

$$(b \times 1) \times \Delta p = 2(L \times 1) \times \tau_0 \qquad \therefore \tau_0 = \frac{b\Delta p}{2L} \tag{7.41}$$

In terms of the friction factor f, defined by

$$f = \frac{\tau_0}{\frac{1}{8}\rho V^2} \tag{7.42}$$

the head loss for the horizontal channel is

$$h_L = \frac{\Delta p}{\gamma} = f\frac{L}{2b}\frac{V^2}{2g} \tag{7.43}$$

Several of the above equations can be combined to find

$$f = \frac{48}{Re} \tag{7.44}$$

where $Re = bV/\nu$. If this is substituted into Eq. (7.43), we see that the head loss is directly proportional to the average velocity in a laminar flow.

The above equations were derived for a channel with an aspect ratio greater than 8. For lower aspect-ratio channels, the sides would require additional terms since the shear acting on the side walls would influence the central part of the flow.

If interest is in a horizontal channel flow where the top plate is moving and there is no pressure gradient, the velocity profile would be the linear profile

$$u(y) = \frac{U}{b}y \tag{7.45}$$

EXAMPLE 7.2
The thin layer of rain at 20°C flows down a parking lot at a relatively constant depth of 4 mm. The area is 40 m wide with a slope of 8 cm over 60 m of length. Estimate (a) the flow rate, (b) shear at the surface, and (c) the Reynolds number.

Solution

(a) The velocity profile can be assumed to be one half of the profile shown in Fig. 7.5, assuming a laminar flow. The average velocity would remain as given by Eq. (7.38), i.e.,

$$V = \frac{b^2 \gamma h}{12 \mu L}$$

where Δp has been replaced with γh. The flow rate is

$$Q = AV = bw \times \frac{b^2 \gamma h}{12 \mu L} = 0.004 \times 40 \times \frac{0.004^2 \times 9810 \times 0.08}{12 \times 10^{-3} \times 60} = 2.80 \times 10^{-3} \, \text{m}^3/\text{s}$$

(b) The shear stress acts only at the solid wall so Eq. (7.41) would provide

$$\tau_0 = \frac{b \gamma h}{L} = \frac{0.004 \times 9810 \times 0.08}{60} = 0.0523 \, \text{Pa}$$

(c) The Reynolds number is

$$\text{Re} = \frac{bV}{v} = \frac{0.004}{10^{-6}} \times \frac{0.004^2 \times 9810 \times 0.08}{12 \times 10^{-3} \times 60} = 69.8$$

The Reynolds number is below 1500 so the assumption of laminar flow is acceptable.

7.4 Laminar Flow Between Rotating Cylinders

Steady flow between concentric cylinders, as shown in Fig. 7.6, is another example of a laminar flow that we can solve analytically. Such a flow exists below a Reynolds number[5] of 1700. Above 1700, the flow might be a complex laminar flow or a turbulent flow. We will again solve this problem using a fluid element and using the appropriate Navier-Stokes equation; either method may be used.

THE ELEMENTAL APPROACH

Two rotating concentric cylinders are displayed in Fig. 7.6. We will assume vertical cylinders so body forces will act normal to the circular flow with the only nonzero velocity component v_θ. The element of fluid selected, shown in Fig. 7.6, has no angular

---

[5]The Reynolds number is defined as $\text{Re} = \omega_1 r_1 \, \delta / v$ where $\delta = r_2 - r_1$.

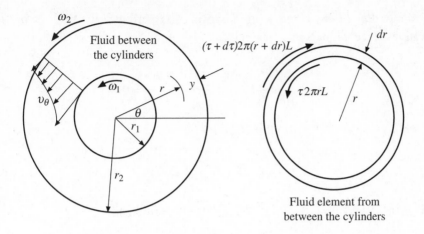

Figure 7.6 Flow between concentric cylinders.

acceleration in this steady-flow condition. Consequently, the summation of torques acting on the element is zero:

$$\tau \times 2\pi r L \times r - (\tau + d\tau) \times 2\pi (r + dr) L \times (r + dr) = 0 \qquad (7.46)$$

where $\tau(r)$ is the shear stress and L is the length of the cylinders, which must be large when compared to the gap width $\delta = r_2 - r_1$. Equation (7.46) simplifies to

$$\tau 2 r dr + r^2 d\tau + 2 r d\tau dr + d\tau (dr)^2 = 0 \qquad (7.47)$$

The last two terms of Eq. (7.46) are higher-order terms that are negligible when compared to the first two terms so that the simplified equation is

$$r \frac{d\tau}{dr} + 2\tau = 0 \qquad (7.48)$$

Now we must recognize that the τ in Eq. (7.48) is[6] $-\tau_{r\theta}$ of Table 5.1 with entry under "Stresses." For this simplified application, the shear stress is related to the velocity gradient by

$$\tau_{r\theta} = \mu r \frac{\partial (v_\theta / r)}{\partial r} \qquad (7.49)$$

---

[6]The minus sign results from the shear stress in Fig.7.6 being on a negative face in the positive direction, the sign convention for a stress component.

This allows Eq. (7.48) to be written, writing the partial derivatives as ordinary derivatives since v_θ depends on r only, as

$$r\mu \frac{d}{dr} r \frac{d(v_\theta/r)}{dr} + 2\mu r \frac{d(v_\theta/r)}{dr} = 0 \qquad (7.50)$$

Multiply by dr, divide by μr, and integrate:

$$r \frac{d(v_\theta/r)}{dr} + 2 \frac{v_\theta}{r} = A \qquad (7.51)$$

or, since $rd/dr(v_\theta/r) = dv_\theta/dr - v_\theta/r$, this can be written as

$$\frac{dv_\theta}{dr} + \frac{v_\theta}{r} = A \qquad \text{or} \qquad \frac{1}{r}\frac{d(rv_\theta)}{dr} = A \qquad (7.52)$$

Now integrate again and obtain

$$v_\theta(r) = \frac{A}{2} r + \frac{B}{r} \qquad (7.53)$$

Using the boundary conditions $v_\theta = r_1\omega_1$ at $r = r_1$ and $v_\theta = r_2\omega_2$ at $r = r_2$, the constants are found to be

$$A = 2\frac{\omega_2 r_2^2 - \omega_1 r_1^2}{r_2^2 - r_1^2} \qquad B = \frac{r_1^2 r_2^2 (\omega_1 - \omega_2)}{r_2^2 - r_1^2} \qquad (7.54)$$

The same result can be obtained by solving the appropriate Navier-Stokes equation; if that is not of interest, skip the next part.

APPLYING THE NAVIER-STOKES EQUATIONS

The θ-component differential momentum equation of Table 5.1 is selected for this circular motion with $v_r = 0$ and $v_z = 0$:

$$\frac{\partial v_\theta}{\partial t} + v_r \frac{\partial v_\theta}{\partial r} + \frac{v_\theta}{r} \frac{\partial v_\theta}{\partial \theta} + v_z \frac{\partial v_\theta}{\partial z} + \frac{v_\theta v_r}{r} = -\frac{1}{\rho r}\frac{\partial p}{\partial \theta} + g_\theta$$

$$+ v\left(\frac{\partial^2 v_\theta}{\partial r^2} + \frac{1}{r}\frac{\partial v_\theta}{\partial r} + \frac{1}{r^2}\frac{\partial^2 v_\theta}{\partial \theta^2} + \frac{\partial^2 v_\theta}{\partial z^2} - \frac{v_\theta}{r^2} + \frac{2}{r^2}\frac{\partial v_r}{\partial \theta} \right) \qquad (7.55)$$

Replace the partial derivatives with ordinary derivatives since v_θ depends on r only and the equation becomes

$$0 = \frac{d^2 v_\theta}{dr^2} + \frac{1}{r}\frac{dv_\theta}{dr} - \frac{v_\theta}{r^2} \qquad (7.56)$$

which can be written in the form

$$\frac{d}{dr}\frac{dv_\theta}{dr} = -\frac{d(v_\theta/r)}{dr} \qquad (7.57)$$

Multiply by dr and integrate:

$$\frac{dv_\theta}{dr} = -\frac{v_\theta}{r} + A \qquad \text{or} \qquad \frac{1}{r}\frac{d(rv_\theta)}{dr} = A \qquad (7.58)$$

Integrate once again:

$$v_\theta(r) = \frac{A}{2}r + \frac{B}{r} \qquad (7.59)$$

The boundary conditions $v_\theta(r_1) = r\omega_1$ and $v_\theta(r_2) = r\omega_2$ allow

$$A = 2\frac{\omega_2 r_2^2 - \omega_1 r_1^2}{r_2^2 - r_1^2} \qquad B = \frac{r_1^2 r_2^2(\omega_1 - \omega_2)}{r_2^2 - r_1^2} \qquad (7.60)$$

QUANTITIES OF INTEREST

Many applications of rotating cylinders involve the outer cylinder being fixed, that is, $\omega_2 = 0$. The velocity distribution, found in the preceding two sections, with A and B simplified, becomes

$$v_\theta(r) = \frac{\omega_1 r_1^2}{r_2^2 - r_1^2}\left(\frac{r_2^2}{r} - r\right) \qquad (7.61)$$

The shear stress τ_1 ($-\tau_{r\theta}$ from Table 5.1) acts on the inner cylinder. It is

$$\tau_1 = -\left[\mu r\frac{d(v_\theta/r)}{dr}\right]_{r=r_1} = \frac{2\mu r_2^2 \omega_1}{r_2^2 - r_1^2} \qquad (7.62)$$

The torque T needed to rotate the inner cylinder is

$$T = \tau_1 A r_1 = \frac{2\mu r_2^2 \omega_1}{r_2^2 - r_1^2} \times 2\pi r_1 L \times r_1 = \frac{4\pi\mu r_1^2 r_2^2 L \omega_1}{r_2^2 - r_1^2} \qquad (7.63)$$

The power $\dot{W}$ required to rotate the inner cylinder with rotational speed ω_1 is

$$\dot{W} = T\omega_1 = \frac{4\pi\mu r_1^2 r_2^2 L \omega_1^2}{r_2^2 - r_1^2} \qquad (7.64)$$

This power, required because of the viscous effects in between the two cylinders, heats up the fluid in bearings and often demands cooling to control the temperature.

For a small gap between the cylinders, as occurs in lubrication problems, it is acceptable to approximate the velocity distribution as a linear profile, a *Couette flow*. Using the variable y of Fig. 7.6 the velocity distribution is

$$v_\theta(r) = \frac{r_1 \omega_1}{\delta} y \qquad (7.65)$$

where y is measured from the outer cylinder in toward the center.

EXAMPLE 7.3

The viscosity is to be determined by rotating a 6-cm-diameter, 30-cm-long cylinder inside a 6.2-cm-diameter cylinder. The torque is measured to be 0.22 N·m and the rotational speed is measured to be 3000 rpm. Use Eqs. (7.61) and (7.65) to estimate the viscosity. Assume that $S = 0.86$.

Solution

The torque is found from Eq. (7.63) based on the velocity distribution of Eq. (7.61):

$$T = \frac{4\pi\mu r_1^2 r_2^2 L \omega_1}{r_2^2 - r_1^2} = \frac{4\pi\mu \times 0.03^2 \times 0.031^2 \times 0.3 \times (3000 \times 2\pi/60)}{0.031^2 - 0.03^2} = 0.22$$

$$\therefore \mu = 0.0131 \text{ N} \cdot \text{s/m}^2$$

Using Eq. (7.65), the torque is found to be

$$T = \tau_1 A r_1 = \mu \frac{r_1 \omega_1}{\delta} \times 2\pi r_1 L \times r_1$$

$$0.22 = \mu \frac{0.03 \times (3000 \times 2\pi/60)}{0.031 - 0.03} \times 2\pi \times 0.03^2 \times 0.3 \qquad \therefore \mu = 0.0138 \text{ N} \cdot \text{s/m}^2$$

The error assuming the linear profile is 5.3 percent.

We should check the Reynolds number to make sure the flow is laminar, as assumed. The Reynolds number is, using $\nu = \mu/\rho$,

$$\text{Re} = \frac{\omega_1 r_1 \delta}{\nu} = \frac{(3000 \times 2\pi/60) \times 0.03 \times 0.001}{0.0131/(1000 \times 0.86)} = 619$$

The laminar flow assumption is acceptable since Re < 1700.

7.5 Turbulent Flow in a Pipe

The Reynolds numbers for most flows of interest in conduits exceed those at which laminar flows cease to exist. If a flow starts from rest, it rather quickly undergoes transition to a turbulent flow. The objective of this section is to express the velocity distribution in a turbulent flow in a pipe and to determine quantities associated with such a flow.

A *turbulent flow* is a flow in which all three velocity components are nonzero and exhibit random behavior; in addition, there must be a correlation between the randomness of at least two of the velocity components. If there is no correlation, it is simply a fluctuating flow. For example, a turbulent boundary layer usually exists near the surface of an airfoil but the flow outside the boundary layer is not referred to as "turbulent" even though there are fluctuations in the flow; it is the free stream.

Let's present one way of describing a turbulent flow. The three velocity components at some point are written as

$$u = \bar{u} + u' \qquad v = \bar{v} + v' \qquad w = \bar{w} + w' \tag{7.66}$$

where $\bar{u}$ denotes a time-average part of the x-component velocity and u' denotes the fluctuating random part. The *time-average* of u is

$$\bar{u} = \frac{1}{T} \int_0^T u(t)\, dt \tag{7.67}$$

where T is sufficiently large when compared to the fluctuation time. For a developed turbulent pipe flow, the three velocity components would appear as in Fig. 7.7. The only time-average component would be $\bar{u}$ in the flow direction. Yet there must exist a correlation between at least two of the random velocity fluctuations, e.g., $\overline{u'v'} \neq 0$; such velocity correlations result in turbulent shear.

We can derive an equation that relates $\overline{u'v'}$ and the time-average velocity component $\bar{u}$ in the flow direction of a turbulent flow but we cannot solve the equation

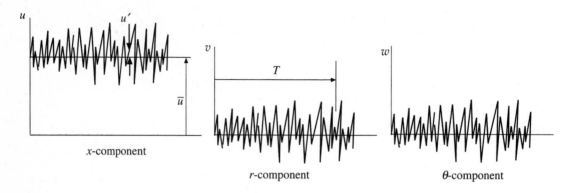

Figure 7.7　The three velocity components in a turbulent flow at a point where the flow is in the x-direction so that $\overline{v} = \overline{w} = 0$ and $\overline{u} \neq 0$.

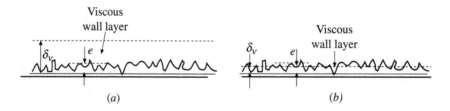

Figure 7.8　(*a*) A smooth wall and (*b*) a rough wall.

even for the simplest case of steady[7] flow in a pipe. So, we will present experimental data for the velocity profile and define some quantities of interest for a turbulent flow in a pipe.

First, let us describe what we mean by a "smooth" wall. A "smooth" wall and a "rough" wall are shown in Fig. 7.8. The *viscous wall layer* is a thin layer near the pipe wall in which the viscous effects are significant. If this viscous layer of thickness δ_v covers the wall roughness elements, the wall is "smooth," as in Fig. 7.8*a*; if the roughness elements protrude out from the viscous layer, the wall is "rough," as in Fig. 7.8*b*.

There are two methods commonly used to describe the turbulent velocity profile in a pipe. These are presented in the following parts.

THE SEMI-LOG PROFILE

The time-average velocity profile in a pipe is presented for a smooth pipe as a semi-log plot in Fig. 7.9 with empirical relationships near the wall and centerline that

---

[7]Steady turbulent flow means the time-average quantities are independent of time.

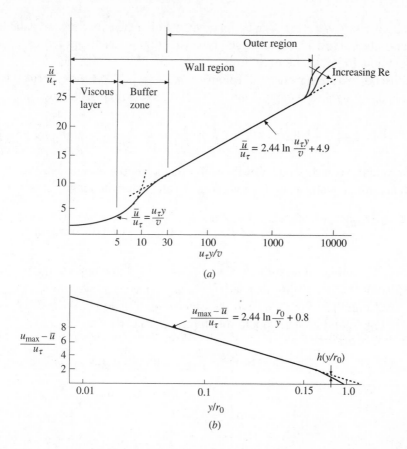

Figure 7.9   Experimental data for a smooth wall in a developed pipe flow. (*a*) The wall region. (*b*) The outer region.

allow $\bar{u}(0) = 0$ and $d\bar{u}/dy = 0$ at $y = r_0$. In the wall region, the characteristic velocity is the *shear velocity*[8] $u_\tau = \sqrt{\tau_0/\rho}$ and the characteristic length is the *viscous length* ν/u_τ; the profiles are

$$\frac{\bar{u}}{u_\tau} = \frac{u_\tau y}{\nu} \qquad 0 \le \frac{u_\tau y}{\nu} \le 5 \qquad \text{(viscous wall layer)} \qquad (7.68)$$

$$\frac{\bar{u}}{u_\tau} = 2.44 \ln \frac{u_\tau y}{\nu} + 4.9 \qquad 30 < \frac{u_\tau y}{\nu} \qquad \frac{y}{r_0} < 0.15 \qquad \text{(turbulent region)} \qquad (7.69)$$

---

[8]The shear velocity is a fictitious velocity that allows experimental data to be presented in dimensionless form that is valid for all turbulent pipe flows. The viscous length is also fictitious.

The interval $5 < u_\tau y/v < 30$ is a *buffer zone* in which the experimental data do not fit either of the curves. The outer edge of the wall region may be as low as $u_\tau y/v = 3000$ for a low Reynolds number flow.

The viscous wall layer plays no role for a rough pipe. The characteristic length is the average roughness height e and the wall region is represented by

$$\frac{\bar{u}}{u_\tau} = 2.44 \ln \frac{y}{e} + 8.5 \qquad \frac{y}{r_0} < 0.15 \qquad \text{(wall region, rough pipe)} \qquad (7.70)$$

The outer region is independent of the wall effects and thus is normalized for both smooth and rough walls using the radius as the characteristic length; it is

$$\frac{u_{max} - \bar{u}}{u_\tau} = -2.44 \ln \frac{y}{r_0} + 0.8 \qquad \frac{y}{r_0} \le 0.15 \qquad \text{(the outer region)} \qquad (7.71)$$

An additional empirical relationship $h(y/r_0)$ is needed to complete the profile for $y > 0.15 r_0$. Most relationships that satisfy $d\bar{u}/dy = 0$ at $y = r_0$ will do.

The wall region of Fig. 7.9a and the outer region of Fig. 7.9b overlap as displayed in Fig. 7.9a. For smooth and rough pipes, respectively,

$$\frac{u_{max}}{u_\tau} = 2.44 \ln \frac{u_\tau r_0}{v} + 5.7 \quad \text{(smooth pipe)} \qquad (7.72)$$

$$\frac{u_{max}}{u_\tau} = 2.44 \ln \frac{r_0}{e} + 9.3 \quad \text{(rough pipes)} \qquad (7.73)$$

We do not often desire the velocity at a particular location but, if we do, before u_{max} can be found, u_τ must be known. To find u_τ we must know τ_0. To find τ_0 we can use [see Eq. (7.5)]

$$\tau_0 = \frac{r_0 \Delta p}{2L} \qquad \text{or} \qquad \tau_0 = \frac{1}{8} \rho V^2 f \qquad (7.74)$$

The friction factor f can be estimated using the power-law profile that follows if the pressure drop is not known.

THE POWER-LAW PROFILE

Another approach, although not quite as accurate as the above, involves the *power-law profile* given by

$$\frac{\bar{u}}{u_{max}} = \left(\frac{y}{r_0} \right)^{1/n} \qquad (7.75)$$

Table 7.1 . Exponent n for Smooth Pipes

Re = VD/v	4×10^3	10^5	10^6	>2×10^6
n	6	7	9	10

where n is between 5 and 10, usually an integer. This can be integrated to give the average velocity:

$$V = \frac{1}{\pi r_0^2} \int_0^{r_0} \bar{u}(r) 2\pi r \, dr = \frac{2n^2}{(n+1)(2n+1)} u_{max} \qquad (7.76)$$

The value of n in Eq. (7.75) is related empirically to f by

$$n = f^{-1/2} \qquad (7.77)$$

For smooth pipes n is related to the Reynolds number in Table 7.1.

The power-law profile cannot be used to estimate the wall shear since it has an infinite slope at the wall for all values of n. It also does not have a zero slope at the pipe centerline so it is not valid near the centerline. It is, however, used to estimate the energy flux and momentum flux of pipe flows.

Finally, it should be noted that the kinetic-energy correction factor is 1.03 if $n = 7$; hence, it is often taken as unity for turbulent flows.

EXAMPLE 7.4
Water at 20°C flows in a 4-cm-diameter pipe with a flow rate of 0.002 m³/s. Estimate (a) the wall shear stress, (b) the maximum velocity, (c) the pressure drop over 20 m, (d) the viscous layer thickness, and (e) determine if the wall is smooth or rough assuming the roughness elements to have a height of 0.0015 mm. Use the power-law profile.

Solution
First, the average velocity and Reynolds number are

$$V = \frac{Q}{A} = \frac{0.002}{\pi \times 0.02^2} = 1.592 \text{ m/s} \qquad Re = \frac{VD}{v} = \frac{1.592 \times 0.04}{10^{-6}} = 6.37 \times 10^4$$

(a) To find the wall shear stress, first find the friction factor. From Table 7.1 the value $n = 6.8$ is selected and from Eq. (7.77)

$$f = \frac{1}{n^2} = \frac{1}{6.8^2} = 0.0216$$

The wall shear stress [see Eq. (7.74)] is

$$\tau_0 = \frac{1}{2}\rho V^2 f = \frac{1}{2} \times 1000 \times 1.592^2 \times 0.0216 = 27.4 \text{ Pa}$$

(b) The maximum velocity is found using Eq. (7.76):

$$u_{max} = \frac{(n+1)(2n+1)}{2n^2}V = \frac{7.8 \times 14.6}{2 \times 6.8^2} \times 1.592 = 1.96 \text{ m/s}$$

(c) The pressure drop is

$$\Delta p = \frac{2L\tau_0}{r_0} = \frac{2 \times 20 \times 27.4}{0.02} = 54\,800 \text{ Pa} \qquad \text{or} \qquad 54.8 \text{ kPa}$$

(d) The friction velocity is

$$u_\tau = \sqrt{\frac{\tau_0}{\rho}} = \sqrt{\frac{27.4}{1000}} = 0.166 \text{ m/s}$$

and the viscous layer thickness is

$$\delta_v = \frac{5\nu}{u_\tau} = \frac{5 \times 10^{-6}}{0.166} = 3.01 \times 10^{-5} \text{ m} \qquad \text{or} \qquad 0.0301 \text{ mm}$$

(e) The height of the roughness elements is given as 0.0015 mm (drawn tubing), which is less than the viscous layer thickness. Hence, the wall is smooth. Note: If the height of the wall elements was 0.046 mm (wrought iron), the wall would be rough.

LOSSES IN PIPE FLOW

The head loss is of considerable interest in pipe flows. It was presented in Eqs. (7.23) and (4.24) and is

$$h_L = f\frac{L}{D}\frac{V^2}{2g} \qquad \text{or} \qquad h_L = \frac{\Delta p}{\gamma} + z_2 - z_1 \qquad (7.78)$$

So, once the friction factor is known, the head loss and pressure drop can be determined. The friction factor depends on a number of properties of the fluid and the pipe:

$$f = f(\rho, \mu, V, D, e) \qquad (7.79)$$

where the roughness height e accounts for the turbulence generated by the roughness elements. A dimensional analysis allows Eq. (7.79) to be written as

$$f = f\left(\frac{e}{D}, \frac{VD\rho}{\mu}\right) \tag{7.80}$$

where e/D is termed the *relative roughness*.

Experimental data has been collected and presented in the form of the *Moody diagram*, displayed in Fig. 7.10 for developed flow in a pipe. The roughness heights are also included on the diagram. There are several features of this diagram that should be emphasized.

- A laminar flow exists up to $\mathrm{Re} \cong 2000$ after which there is a *critical zone* in which the flow is undergoing transition to a turbulent flow. This may involve transitory flow that alternates between laminar and turbulent flows.

- The friction factor in the *transition zone*, which begins at about $\mathrm{Re} = 4000$ and decreases with increasing Reynolds numbers, becomes constant at the end of the zone, as signified by the dashed line.

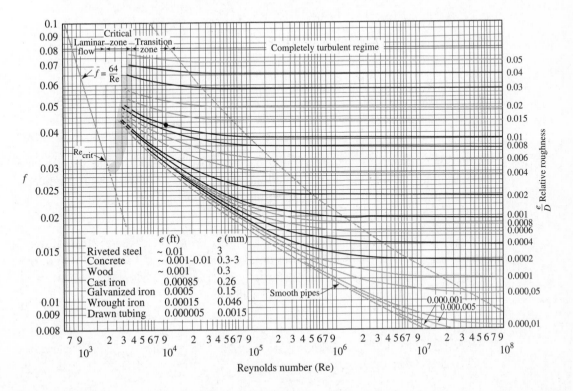

Figure 7.10   The Moody diagram.

- The friction factor in the *completely turbulent zone* is constant and depends on the relative roughness e/D. Viscous effects, and thus the Reynolds number, do not affect the friction factor.

- The height e of the roughness elements in the Moody diagram is for new pipes. Pipes become fouled with age changing both e and the diameter D resulting in an increased friction factor. Designs of piping systems should include such aging effects.

An alternate to using the Moody diagram is to use formulas developed by Swamee and Jain for pipe flow; the particular formula selected depends on the information given. The formulas to determine quantities in long reaches of developed pipe flow (these formulas are not used in short lengths or in pipes with numerous fittings and geometry changes) are as follows:

$$h_L = 1.07 \frac{Q^2 L}{gD^5} \left\{ \ln\left[\frac{e}{3.7D} + 4.62\left(\frac{vD}{Q}\right)^{0.9} \right] \right\}^{-2} \qquad \begin{array}{l} 10^{-6} < \dfrac{e}{D} < 10^{-2} \\[2mm] 3000 < \mathrm{Re} < 3\times10^8 \end{array} \qquad (7.81)$$

$$Q = -0.965 \sqrt{\frac{gD^5 h_L}{L}} \ \ln\left[\frac{e}{3.7D} + \left(\frac{3.17v^2 L}{gD^3 h_L}\right)^{0.5} \right] \qquad 2000 < \mathrm{Re} \qquad (7.82)$$

$$D = 0.66 \left[e^{1.25}\left(\frac{LQ^2}{gh_L}\right)^{4.75} + vQ^{9.4}\left(\frac{L}{gh_L}\right)^{5.2} \right]^{0.04} \qquad \begin{array}{l} 10^{-6} < \dfrac{e}{D} < 10^{-2} \\[2mm] 5000 < \mathrm{Re} < 3\times10^8 \end{array} \qquad (7.83)$$

Either SI or English units can be used in the above equations. Note also that the Moody diagram and the above equations are accurate to within about 5 percent, sufficiently accurate for most engineering applications.

EXAMPLE 7.5

A pressure drop of 500 kPa is measured over 200 m of a horizontal length of 8-cm-diameter cast iron pipe transporting water at 20°C. Estimate the flow rate using (a) the Moody diagram and (b) an alternate equation.

Solution

The relative roughness (find e in Fig. 7.10) is

$$\frac{e}{D} = \frac{0.26}{80} = 0.00325$$

Assuming a completely turbulent flow, the friction factor from Fig. 7.10 is $f = 0.026$. The head loss is

$$h_L = \frac{\Delta p}{\gamma} = \frac{500\,000}{9800} = 51 \text{ m}$$

The average velocity, from Eq. (7.78) is

$$V = \sqrt{\frac{2gDh_L}{fL}} = \sqrt{\frac{2 \times 9.8 \times 0.08 \times 51}{0.026 \times 200}} = 3.92 \text{ m/s}$$

We must check the Reynolds number to make sure the flow is completely turbulent. It is

$$\text{Re} = \frac{VD}{v} = \frac{3.92 \times 0.08}{10^{-6}} = 3.14 \times 10^5$$

This is just acceptable and requires no iteration to improve the friction factor. So, the flow rate is

$$Q = AV = \pi \times 0.04^2 \times 3.92 = 0.0197 \text{ m}^3/\text{s}$$

(b) Use the alternate equation that relates Q to the other quantities, i.e., Eq. (7.82). We use the head loss from part (a):

$$Q = -0.965\sqrt{\frac{9.8 \times .08^5 \times 51}{200}}\ln\left[\frac{0.26}{3.7 \times 80} + \left(\frac{3.17 \times 10^{-12} \times 200}{9.8 \times 0.08^3 \times 51}\right)^{0.5}\right] = 0.0193 \text{ m}^3/\text{s}$$

This equation was easier to use and gave an acceptable result.

LOSSES IN NONCIRCULAR CONDUITS

To determine the head loss in a relatively "open" noncircular conduit, we use the *hydraulic radius R*, defined as

$$R = \frac{A}{P} \tag{7.84}$$

where A is the cross-sectional area and P is the *wetted perimeter,* the perimeter of the conduit that is in contact with the fluid. The Reynolds number, relative roughness, and head loss are, respectively,

$$\text{Re} = \frac{4VR}{v} \qquad \text{relative roughness} = \frac{e}{4R} \qquad h_L = f\frac{L}{4R}\frac{V^2}{2g} \tag{7.85}$$

A rectangular area should have an aspect ratio less than 4. This method should not be used with more complex shapes, like that of an annulus.

MINOR LOSSES

The preceding losses were for the developed flow in long conduits. Most piping systems, however, include sudden changes such as elbows, valves, inlets, etc., that add additional losses. These losses are called *minor losses* that may, in fact, add up to exceed the head loss found in the preceding sections. These minor losses are expressed in terms of a *loss coefficient K*, defined for most devices by

$$h_L = K\frac{V^2}{2g} \tag{7.86}$$

A number of loss coefficients are included in Table 7.2. Note that relatively low loss coefficients are associated with gradual contractions whereas relatively large coefficients with enlargements, due to the separated flows in enlargements. Separated and secondary flows also occur in elbows resulting in surprisingly large loss coefficients. Vanes that eliminate such separated or secondary flows can substantially reduce the losses, as noted in the table.

We often equate the losses in a device to an *equivalent length* of pipe, i.e.,

$$h_L = K\frac{V^2}{2g} = f\frac{L_e}{D}\frac{V^2}{2g} \tag{7.87}$$

This provides the relationship

$$L_e = K\frac{D}{f} \tag{7.88}$$

If a pipe is quite long, greater than 1000 diameters, minor losses can be neglected. For lengths as short as 100 diameters, the minor losses may exceed the frictional losses. For short and intermediate lengths, the minor losses should be included.

Table 7.2 Minor Loss Coefficients K for Selected Devices*

Type of fitting diameter	Screwed			Flanged		
	2.5 cm	5 in.	10 cm	5 cm	10 cm	20 cm
Globe value (fully open)	8.2	6.9	5.7	8.5	6.0	5.8
(half open)	20	17	14	21	15	14
(one-quarter open)	57	48	40	60	42	41
Angle valve (fully open)	4.7	2.0	1.0	2.4	2.0	2.0
Swing check valve (fully open)	2.9	2.1	2.0	2.0	2.0	2.0
Gate valve (fully open)	0.24	0.16	0.11	0.35	0.16	0.07
Return bend	1.5	0.95	0.64	0.35	0.30	0.25
Tee (branch)	1.8	1.4	1.1	0.80	0.64	0.58
Tee (line)	0.9	0.9	0.9	0.19	0.14	0.10
Standard elbow	1.5	0.95	0.64	0.39	0.30	0.26
Long sweep elbow	0.72	0.41	0.23	0.30	0.19	0.15
45° elbow	0.32	0.30	0.29			
Square-edged entrance			0.5			
Reentrant entrance			0.8			
Well-rounded entrance			0.03			
Pipe exit			1.0			
Sudden contraction†	Area ratio					
	2:1		0.25			
	5:1		0.41			
	10:1		0.46			
Orifice plate	Area ratio A/A_0					
	1.5:1		0.85			
	2:1		3.4			
	4:1		29			
	≥6:1		$2.78\left(\dfrac{A}{A_0}-0.6\right)^2$			
Sudden enlargement‡			$\left(1-\dfrac{A_1}{A_2}\right)^2$			
90° miter bend (without vanes)			1.1			
(with vanes)			0.2			
General contraction (30° included angle)			0.02			
(70° included angle)			0.07			

*Values for other geometries can be found in *Technical Paper 410*. The Crane Company, 1957.

†Based on exit velocity V_2.

‡Based on entrance velocity V_1.

EXAMPLE 7.6

A 1.5-cm-diameter, 20-m-long plastic pipe transports water from a pressurized 400-kPa tank out to a free open end located 3 m above the water surface in the tank. There are three elbows in the water line and a square-edged inlet from the tank. Estimate the flow rate.

Solution

The energy equation is applied between the tank and the faucet exit:

$$0 = \frac{V_2^2 - \cancel{V_1^2}}{2g} + \frac{\cancel{p_2} - p_1}{\gamma} + z_2 - z_1 + h_L$$

where

$$h_L = \left(f \frac{L}{D} + 3 \times K_{\text{elbow}} + K_{\text{entrance}} \right) \frac{V^2}{2g}$$

Assume that the pipe has $e/D = 0$ and that $\text{Re} \cong 2 \times 10^5$ so that the Moody diagram gives $f = 0.016$. The energy equation gives

$$0 = \frac{V_2^2}{2 \times 9.8} - \frac{400\,000}{9800} + 3 + \left(0.016 \times \frac{20}{0.015} + 3 \times 1.6 + .5 \right) \times \frac{V^2}{2 \times 9.8}$$

$$\therefore V = 5.18 \text{ m/s}$$

The Reynolds number is then $\text{Re} = 5.18 \times 0.15/10^{-6} = 7.8 \times 10^4$. Try $f = 0.018$:

$$0 = \frac{V_2^2}{2 \times 9.8} - \frac{400\,000}{9800} + 3 + \left(0.018 \times \frac{20}{0.015} + 3 \times 1.6 + .5 \right) \times \frac{V^2}{2 \times 9.8}$$

$$\therefore V = 4.95 \text{ m/s}$$

resulting in $\text{Re} = 4.95 \times 0.15/10^{-6} = 7.4 \times 10^4$. This is close enough, so we use $V = 5.0$ m/s. The flow rate is

$$Q = AV = \pi \times 0.0075^2 \times 5 = 8.8 \times 10^{-4} \text{ m}^3/\text{s}$$

HYDRAULIC AND ENERGY GRADE LINES

The energy equation is most often written so that each term has dimensions of length, i.e.,

$$-\frac{\dot{W}_S}{\dot{m}g} = \frac{V_2^2 - V_1^2}{2g} + \frac{p_2 - p_1}{\gamma} + z_2 - z_1 + h_L \qquad (7.89)$$

In piping systems it is often conventional to refer to the hydraulic grade line and the energy grade line. The *hydraulic grade line* (HGL), the dashed line in Fig. 7.11, is the locus of points located a distance p/γ above the centerline of a pipe. The *energy grade line* (EGL), the solid line in Fig. 7.11, is the locus of points located a distance $V^2/2$ above the HGL. The following observations relate to the HGL and the EGL.

- The EGL approaches the HGL as the velocity goes to zero. They are identical on the surface of a reservoir.

- Both the EGL and the HGL slope downward in the direction of the flow due to the losses in the pipe.

- A sudden drop occurs in the EGL and the HGL equal to the loss due to a sudden geometry change, such as an entrance or a valve.

- A jump occurs in the EGL and the HGL due to a pump and a drop due to a turbine.

- If the HGL is below the pipe, there is a vacuum in the pipe, a condition that is most often avoided in piping systems because of possible contamination.

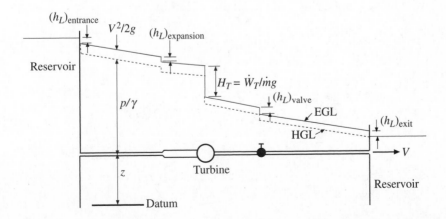

Figure 7.11   The hydraulic grade line (HGL) and the energy grade line (EGL).

7.6 Open Channel Flow

Consider the developed turbulent flow in an open channel, shown in Fig. 7.12. The water flows at a depth of y and the channel is on a slope S, which is assumed to be small so that $\sin\theta = S$. The cross section could be trapezoidal, as shown, or it could be circular or triangular. Let us apply the energy equation between the two sections:

$$0 = \frac{V_2^2 - V_1^2}{2g} + \frac{p_2 - p_1}{\gamma} + z_2 - z_1 + h_L \tag{7.90}$$

The head loss is the elevation change, i.e.,

$$h_L = z_1 - z_2 = L\sin\theta = LS \tag{7.91}$$

where L is the distance between the two selected sections. Using the head loss expressed by Eq. (7.85), we have

$$h_L = f\frac{L}{4R}\frac{V^2}{2g} = LS \quad \text{or} \quad V^2 = \frac{8g}{f}RS \tag{7.92}$$

The Reynolds number of the flow in an open channel is invariably large and the channel rough so that the friction factor is a constant independent of the velocity (see the Moody diagram of Fig. 7.10) for a particular channel. Consequently, the velocity is related to the slope and hydraulic radius by

$$V = C\sqrt{RS} \tag{7.93}$$

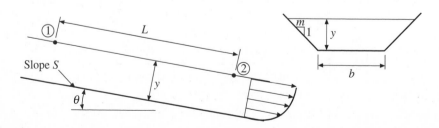

Figure 7.12   Flow in an open channel.

Table 7.3 Values* of the Manning n

Wall material	Manning n
Brick	0.016
Cast or wrought iron	0.015
Concrete pipe	0.015
Corrugated metal	0.025
Earth	0.022
Earth with stones and weeds	0.035
Finished concrete	0.012
Mountain streams	0.05
Planed wood	0.012
Sewer pipe	0.013
Riveted steel	0.017
Rubble	0.03
Unfinished concrete	0.014
Rough wood	0.013

*The values in this table result in flow rates too large for $R > 3$ m. The Manning n should be increased by 10 to 15 percent for the larger channels.

where C is a dimensional constant called the *Chezy coefficient*; it has been related experimentally to the channel roughness and the hydraulic radius by

$$C = \frac{1}{n} R^{1/6} \tag{7.94}$$

The dimensionless constant n is a measure of the wall roughness and is called the *Manning n*. Values for a variety of wall materials are listed in Table 7.3.

The flow rate in an open channel follows from $Q = AV$ and is

$$Q = \frac{1}{n} A R^{2/3} S^{1/2} \tag{7.95}$$

This is referred to as the *Chezy-Manning equation*. It can be applied using English units by replacing the "1" in the numerator with "1.49."

If the channel surface is smooth, e.g., glass or plastic, Eq. (7.95) should not be used since it assumes a rough surface. For channels with smooth surfaces the Darcy-Weisbach equation, Eq. (7.85), along with the Moody diagram should be used.

EXAMPLE 7.7

Water at 20°C is flowing in a 2-m-wide rectangular, brick channel at a depth of 120 cm. The slope is 0.0012. Estimate the flow rate using (a) the Chezy-Manning equation and (b) the Darcy-Weisbach equation.

Solution

First, calculate the hydraulic radius:

$$R = \frac{A}{P} = \frac{by}{b+2y} = \frac{2 \times 1.2}{2 + 2 \times 1.2} = 0.545 \text{ m}$$

(a) The Chezy-Manning equation provides

$$Q = \frac{1}{n} AR^{2/3} S^{1/2}$$

$$= \frac{1}{0.016} \times (2 \times 1.2) \times 0.545^{2/3} \times 0.0012^{1/2} = 3.47 \text{ m}^3/\text{s}$$

(b) To use the Darcy-Weisbach equation, we must find the friction factor f. The Moody diagram requires a value for e. Use a relatively large value such as that for rougher concrete, i.e., $e = 2$ mm. Since the hydraulic radius $R = D/4$ for a circle, we use

$$\frac{e}{D} = \frac{e}{4R} = \frac{0.002}{4 \times 0.545} = 0.00092$$

The Moody diagram gives $f \cong 0.019$. The Darcy-Weisbach equation takes the form of Eq. (7.92):

$$V = \sqrt{\frac{8g}{f} RS} = \sqrt{\frac{8 \times 9.8}{0.019} \times 0.545 \times 0.0012} = 1.64 \text{ m/s}$$

The flow rate is then

$$Q = AV = 2 \times 1.2 \times 1.64 = 3.94 \text{ m}^3/\text{s}$$

Check the Reynolds number:

$$\text{Re} = \frac{4VR}{v} = \frac{4 \times 1.64 \times 0.545}{10^{-6}} = 3.6 \times 10^{6}$$

This is sufficiently large so that f is acceptable. Note that the Q of part (a) is about 12 percent lower than that of part (b). That of part (b) is considered more accurate.

Quiz No. 1

1. The maximum average velocity for a laminar flow of SAE-30 oil at 80°C in a 2-cm-diameter pipe using a critical Reynolds number of 2000 is nearest

    (A) 1.8 m/s

    (B) 1.4 m/s

    (C) 1.0 m/s

    (D) 0.82 m/s

2. Water is flowing in a 2-cm-diameter pipe with a flow rate of 0.0002 m³/s. For a conventional entrance, estimate the entrance length if the water temperature is 40°C.

    (A) 10 m

    (B) 15 m

    (C) 20 m

    (D) 25 m

3. Water at 40°C flows through a 4-cm-diameter pipe at a rate of 6 L/min. Assuming a laminar flow, the pressure drop over 20 m of length in the horizontal pipe is nearest

    (A) 11 Pa

    (B) 21 Pa

    (C) 35 Pa

    (D) 42 Pa

4. Water at 20°C flows through a 12-mm-diameter pipe on a downward slope so that Re = 2000. What angle would result in a zero pressure drop?

    (A) 0.86°

    (B) 0.48°

    (C) 0.22°

    (D) 0.15°

5. What pressure gradient dp/dx would provide a zero shear stress on the stationary lower plate for horizontal plates with the top plate moving to the right with velocity U. Assume a laminar flow with b separating the plates.

    (A) U/b^2

    (B) $2U/b^2$

    (C) $4U/b^2$

    (D) $U/2b^2$

6. Assuming a Couette flow between a stationary and a rotating cylinder, the power needed to rotate the inner rotating cylinder is

    (A) $2\pi\mu r_1^3 \omega^2 L/\delta$

    (B) $\pi\mu r_1^3 \omega^2 L^2/\delta^2$

    (C) $\pi\mu r_1^3 \omega^2 L/\delta$

    (D) $2\pi\mu r_1^3 \omega^2 L^2/\delta^2$

7. A 12-cm-diameter pipe transports water at 25°C in a pipe with roughness elements averaging 0.26 mm in height. Estimate the maximum velocity in the pipe if the flow rate is 0.0004 m³/s.

    (A) 0.075 m/s

    (B) 0.065 m/s

    (C) 0.055 m/s

    (D) 0.045 m/s

8. SAE-30 oil at 20°C is transported in a smooth 40-cm-diameter pipe with an average velocity of 10 m/s. Using the power-law velocity profile, estimate the pressure drop over 100 m of pipe.

    (A) 300 kPa

    (B) 275 kPa

    (C) 250 kPa

    (D) 225 kPa

9. Water at 20°C flows at 0.02 m³/s in an 8-cm-diameter galvanized-iron pipe. Calculate the head loss over 40 m of horizontal pipe

   (A) 6.9 m

   (B) 8.1 m

   (C) 9.7 m

   (D) 10.3 m

10. If the pressure drop in a 100-m section of horizontal 10-cm-diameter galvanized-iron pipe is 200 kPa, estimate the flow rate if water at 20°C is flowing.

   (A) 0.018 m³/s

   (B) 0.027 m³/s

   (C) 0.033 m³/s

   (D) 0.041 m³/s

11. Water at 20°C flows from a reservoir out of a 100-m-long, 4-cm-diameter galvanized-iron pipe to the atmosphere. The outlet is 20 m below the surface of the reservoir. What is the exit velocity? There is a square-edged entrance.

   (A) 2.0 m/s

   (B) 3.0 m/s

   (C) 4.0 m/s

   (D) 5.0 m/s

12. Water flows in a 2-m-wide rectangular finished concrete channel with a slope of 0.001 at a depth of 80 cm. Estimate the flow rate.

   (A) 2.15 m³/s

   (B) 2.45 m³/s

   (C) 2.75 m³/s

   (D) 2.95 m³/s

Quiz No. 2

1. A drinking fountain has an opening of 4 mm in diameter. The water rises a distance of about 20 cm in the air. Is the flow laminar or turbulent as it leaves the opening? Make any assumptions needed.

2. SAE-30 oil at 80°C occupies the space between two cylinders of 2 cm and 2.2 cm in diameter. The outer cylinder is stationary and the inner cylinder rotates at 1000 rpm. Is the oil in a laminar or turbulent state? Use $\text{Re}_{crit} = 1700$ where $\text{Re} = \omega r_1 \delta / \nu$ and $\delta = r_2 - r_1$.

3. A parabolic velocity profile is desired at the end of a 10-m-long, 8-mm-diameter tube attached to a tank filled with water 20°C. An experiment is run during which 60 L is collected in 90 min. Is the laminar flow assumption reasonable?

4. The pressure drop over a 15-m length of 8-mm-diameter horizontal pipe transporting water at 40°C is measured to be 1200 Pa. A laminar flow is assumed. Determine the wall shear stress and the friction factor.

5. What pressure gradient is needed so that the flow rate is zero for laminar flow between horizontal parallel plates if the lower plate is stationary and the top plate moves with velocity U? The distance b separates the plates.

6. Water at 20°C flows down an 80-m-wide parking lot at a constant depth of 5 mm. The slope of the parking lot is 0.0002. Estimate the flow rate and the maximum shear stress.

7. SAE-10 oil at 20°C fills the gap between a rotating cylinder and a fixed outer cylinder. Estimate the torque needed to rotate a 20-cm-long cylinder at 40 rad/s assuming a Couette flow.

8. A 12-cm-diameter pipe transports water at 25°C in a pipe with roughness elements averaging 0.26 mm in height. Decide if the pipe is smooth or rough if the flow rate is 0.0004 m^3/s.

9. SAE-30 oil at 20°C is transported in a smooth 40-cm-diameter pipe with an average velocity of 10 m/s. Using the power-law velocity profile, estimate the viscous wall layer thickness.

10. SAE-10 oil at 80°C flows at 0.02 m^3/s in an 8-cm-diameter galvanized-iron pipe. Calculate the head loss over 40 m of horizontal pipe.

11. A pressure drop of 6000 Pa is measured over a 20 m length as water at 30°C flows through the 2- by 6-cm smooth conduit. Estimate the flow rate.

12. An 88 percent efficient pump is used to transport 30°C water from a lower reservoir through a 8-cm-diameter galvanized-iron pipe to a higher reservoir whose surface is 40 m above the surface of the lower one. The pipe has a total length of 200 m. Estimate the power required for a flow rate of 0.04 m^3/s.

13. Water is not to exceed a depth of 120 cm in a 2-m-wide finished concrete channel on a slope of 0.001. What would the flow rate be at that depth?

CHAPTER 8

External Flows

The subject of external flows involves both low Reynolds-number flows and high Reynolds-number flows. Low Reynolds-number flows are not of interest in most engineering applications and will not be considered; flow around spray droplets, river sediment, filaments, and red blood cells would be examples that are left to the specialists. High Reynolds-number flows, however, are of interest to many engineers and include flow around airfoils, vehicles, buildings, bridge cables, stadiums, turbine blades, and signs, to name a few.

It is quite difficult to solve for the flow field external to a body, even the simplest of bodies like a long cylinder or a sphere. We can, however, develop equations that allow us to estimate the growth of the thin viscous layer, the boundary layer, which grows on a flat plate or the rounded nose of a vehicle. Also, coefficients have been determined experimentally that allow the drag and the lift to be approximated with sufficient accuracy.

8.1 Basics

The flow around a blunt body involves a *separated region*, a region in which the flow separates from the body and forms a recirculating region downstream, as

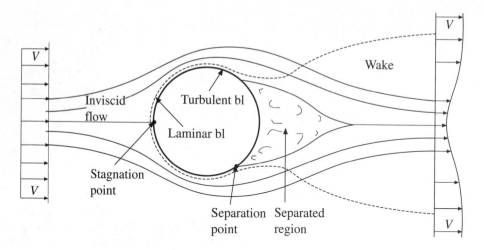

Figure 8.1 The details of a flow around a blunt body.

shown in Fig. 8.1. A *wake*, a region influenced by viscosity, is also formed; it is a diffusive region that continues to grow (some distance downstream the velocity is less than the free-stream velocity V). A laminar boundary layer exists near the front of the body followed by a turbulent boundary layer as shown in Fig. 8.1. An inviscid flow, often referred to as the *free stream*, exists on the front of the body and outside the boundary layer, separated region, and wake. The flow around a streamlined body has all the same components as that of Fig. 8.1 except it does not have a significant separated region, and the wake is much smaller.

The free-stream inviscid flow is usually irrotational although it could be a rotational flow with vorticity, e.g., the flow of air near the ground around a tree trunk or water near the ground around a post in a river; the water digs a depression in the sand in front of the post and the air digs a similar depression in snow in front of the tree, a rather interesting observation. The vorticity in the approaching air or water accounts for the observed phenomenon.

It should be noted that the boundary of the separated region is shown at an average location. It is, however, highly unsteady and is able to slowly exchange mass with the free stream, even though the time-average streamlines remain outside the separated region. Also, the separated region is always located inside the wake.

Interest in the flow around a blunt object is focused on the *drag*, the force the flow exerts on the body in the direction of the flow.[1] *Lift* is the force exerted normal

---

[1]Actually, the body moves through the stationary fluid. To create a steady flow, the fluid moves past the stationary body, as in a laboratory wind tunnel; pressures and forces remain the same. To obtain the actual velocity, the flow velocity through the wind tunnel is subtracted from the velocity at each point.

to the flow direction and is of interest on airfoils and streamlined bodies. The drag F_D and lift F_L are specified in terms of the *drag coefficient C_D* and *lift coefficient C_L*, respectively, by

$$F_D = \frac{1}{2}\rho A V^2 C_D \quad \text{and} \quad F_L = \frac{1}{2}\rho A V^2 C_L \tag{8.1}$$

where, for a blunt body, the area A is the area projected on a plane normal to the flow direction, and for an airfoil the area A is the *chord* (the distance from the nose to the trailing edge) times the length.

The force due to the lower pressure in the separated region dominates the drag force on a blunt body, the subject of Sec. 8.2. The viscous stress that acts on and parallel to each boundary element is negligible and thus little, if any, attention is paid to the boundary layer on the surface of a blunt body. The opposite is true for an airfoil, the subject of Sec. 8.3; the drag force is primarily due to the viscous stresses that act on the boundary elements. Consequently, there is considerable interest in the boundary layer that develops on a streamlined body.

The basics of boundary-layer theory will be presented in Sec. 8.5. But first, the inviscid flow outside the boundary layer (see Fig. 8.1) must be known. So, inviscid flow theory will be presented in Sec. 8.4. The boundary layer is so thin that it can be ignored when solving for the inviscid flow. The inviscid flow solution provides the lift, which is not significantly influenced by the viscous boundary layer, and it also provides the pressure distribution on the body's surface as well as the velocity on that surface (since the inviscid solution ignores the effects of viscosity, the fluid does not stick to the boundary but slips by the boundary). Both the pressure and the velocity at the surface are needed in the boundary-layer solution.

8.2 Flow Around Blunt Bodies

DRAG COEFFICIENTS

The primary flow parameter that influences the drag around a blunt body is the Reynolds number. If there is no free surface, the drag coefficients for both smooth and rough spheres and long cylinders are presented in Fig. 8.2; the values for streamlined cylinders and spheres are also included.

Separation always occurs in the flow of a fluid around a blunt body if the Reynolds number is sufficiently high. However, at low Reynolds numbers (it is

Fluid Mechanics Demystified

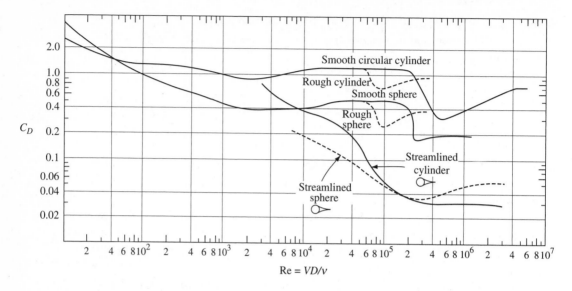

Figure 8.2 Drag coefficients for flow around spheres and long cylinders (E. Achenbach, J. *"Fluid Mech.,"* v.54, 1972).

called a *Stokes flow* if Re < 5), there is no separation and the drag coefficient, for a sphere, is given by

$$C_D = \frac{24}{\text{Re}} \qquad \text{Re} < 1 \tag{8.2}$$

Separation occurs for Re ≥ 10 beginning over a small area on the rear of the sphere until the separated region reaches a maximum at Re ≅ 1000. The drag coefficient is then relatively constant until a sudden drop occurs in the vicinity of Re = 2 × 10⁵. This sudden drop is due to the transition of the boundary layer just before separation undergoing transition from a laminar flow to a turbulent flow. A turbulent boundary layer contains substantially more momentum and is able to move the separation region further to the rear (see the comparison in Fig. 8.3). The sudden decrease in drag could be as much as 80 percent. The surface of an object can be roughened to cause the boundary layer to undergo transition prematurely; the dimples on a golf ball accomplish this and increase the flight by up to 100 percent when compared to the flight of a smooth ball.

After the sudden drop occurs, the drag coefficient again increases with increased Reynolds number. Experimental data does not provide the drag coefficients for either the sphere or the cylinder for high Reynolds numbers. The values of 0.4 for

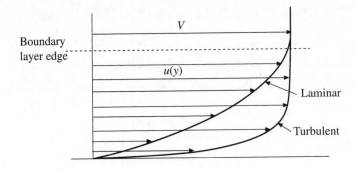

Figure 8.3 Laminar and turbulent velocity profiles for the same boundary
layer thickness.

long smooth cylinders and 0.2 for smooth spheres for Reynolds numbers exceeding
10^6 are often used (contrary to the data of Fig. 8.2).

Streamlining can substantially reduce the drag coefficients of blunt bodies. The
drag coefficients for streamlined cylinders and spheres are also shown in Fig. 8.2.
The included angle at the trailing edge should not exceed about 20° if the separated
region is to be minimized. The drag due to the shear stress acting on the enlarged
surface will certainly increase for a streamlined body, but the drag due to the low
pressure will be reduced much more so that the total drag will be less. Also, streamlining eliminates the vortices that cause vibrations when shed from a blunt body.

For cylinders of finite length with free ends, the drag coefficient must be reduced
using the data of Table 8.1. If a finite-length cylinder has one end fixed to a solid

Table 8.1 Drag Coefficients for Finite-Length Circular
Cylinders* with Free Ends†

L/D	$C_D/C_{D\infty}$
∞	1
40	0.82
20	0.76
10	0.68
5	0.62
3	0.62
2	0.57
1	0.53

*The drag coefficient $C_{D\infty}$ is from Fig. 8.2.
†If one end is fixed to a solid surface, double the length of the cylinder.

Table 8.2   Drag Coefficients for Various Blunt Objects

Object	Re	C_D
Square cylinder of width w		
$L/w = \begin{cases} \infty \\ 1 \end{cases}$	$>10^4$ $>10^4$	2.0 1.1
Rectangular plates		
$L/w = \begin{cases} \infty \\ 20 \\ 5 \\ 1 \end{cases}$	$>10^3$ $>10^3$ $>10^3$ $>10^3$	2.0 1.5 1.2 1.1
Circular disc	$>10^3$	1.1
Parachute	$>10^7$	1.4
Modern automobile	$>10^5$	0.29
Van	$>10^5$	0.42
Bicycle $\begin{cases} \text{upright rider} \\ \text{bent over rider} \\ \text{drafting rider} \end{cases}$		1.1 0.9 0.5
Semitruck $\begin{cases} \text{standard} \\ \text{with streamlined deflector} \\ \text{with deflector and gap seal} \end{cases}$		0.96 0.76 0.07

surface, the length of the cylinder is doubled. Note that the L/D of a cylinder with free ends has to be quite large before the end effects are not significant.

Drag coefficients for a number of common shapes that are insensitive to high Reynolds numbers are presented in Table 8.2.

EXAMPLE 8.1

A 5-cm-diameter, 6-m-high pole fixed in concrete supports a flat, circular 4-m-diameter sign. For a wind speed of 30 m/s, estimate the maximum moment that must be resisted by the concrete.

Solution

To obtain the maximum moment, the wind is assumed normal to the sign. From Table 8.2 the drag coefficient for a disc is 1.1. The moment due to the drag force, which acts at the center of the sign, is

$$M_1 = F_{D1} \times L_1 = \frac{1}{2}\rho A_1 V^2 C_{D1} \times L_1$$

$$= \frac{1}{2} \times 1.22 \times \pi \times 2^2 \times 30^2 \times 1.1 \times 8 = 60\,700 \text{ N} \cdot \text{m}$$

where the density at sea level of 1.22 kg/m³ is used since the elevation is not given. The moment due to the pole is

$$M_2 = F_{D2} \times L_2 = \frac{1}{2}\rho A_2 V^2 C_{D2} \times L_2$$

$$= \frac{1}{2} \times 1.22 \times 0.05 \times 6 \times 30^2 \times 0.7 \times 3 = 346 \text{ N} \cdot \text{m}$$

assuming a Reynolds number of $\text{Re} = 30 \times 0.05 / 1.5 \times 10^{-5} = 10^{-5}$ and high-intensity fluctuations in the air flow, i.e., a rough cylinder. The factor from Table 8.1 was not used since neither end was free. The moment that must be resisted by the concrete base is

$$M = M_1 + M_2 = 60\,700 + 346 = 61\,000 \text{ N} \cdot \text{m}$$

VORTEX SHEDDING

Long cylindrical bodies exposed to a fluid flow can exhibit the phenomenon of *vortex shedding* at relatively low Reynolds numbers. Vortices are shed from electrical wires, bridges, towers, and underwater communication wires, and can actually experience significant damage. We will consider the vortices shed from a long circular cylinder. The shedding occurs alternately from each side of the cylinder, as shown in Fig. 8.4. The shedding frequency f, in hertz, is given by the Strouhal number,

$$\text{St} = \frac{fD}{V} \tag{8.3}$$

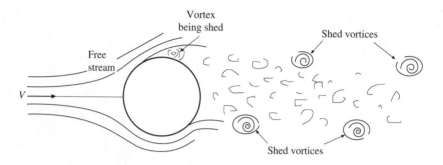

Figure 8.4 Vortices shed from a cylinder.

If this shedding frequency is the same, or a multiple of a structure's frequency, then there is the possibility that damage may occur due to resonance.

The shedding frequency cannot be calculated from equations; it is determined experimentally and shown in Fig. 8.5. Note that vortex shedding initiates at Re ≈ 40 and for Re ≥ 300 the Strouhal number is essentially independent of Reynolds number and is equal to about 0.21. The vortex-shedding phenomenon disappears for Re > 10^4.

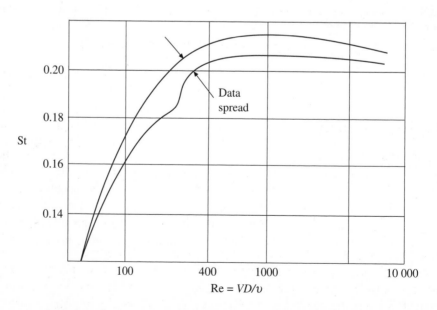

Figure 8.5 Strouhal number for vortex shedding from a cylinder.

EXAMPLE 8.2
A 6-cm-diameter cylinder is used to measure the velocity of a slow-moving air stream. Two pressure taps are used to determine that the vortices are shed with a frequency of 4 Hz. Determine the velocity of the air stream.

Solution
Assume the Strouhal number to be in the range $300 < \text{Re} < 10\ 000$. Then

$$\frac{fD}{V} = 0.21 \qquad \text{so that} \qquad V = \frac{4 \times 0.06}{0.21} = 1.14 \text{ m/s}$$

It is quite difficult to measure the velocity of an air stream this low. The measurement of the shed vortices is one method of doing so.

CAVITATION

When a liquid flows from a region of relatively high pressure into a region of low pressure, *cavitation* may occur, that is, the pressure may be sufficiently low so that the liquid vaporizes. This can occur in pipe flows in which a contraction and expansion exists: in the vanes of a centrifugal pump, near the tips of propellers, on hydrofoils, and torpedoes. It can actually damage the propellers and the steel shafts (due to vibrations) on ships and cause a pump to cease to function properly. It can, however, also be useful in the destruction of kidney stones, in ultrasonic cleaning devices, and in improving the performance of torpedoes.

Cavitation occurs whenever the *cavitation number* σ, defined by

$$\sigma = \frac{p_\infty - p_v}{\frac{1}{2}\rho V^2} \tag{8.4}$$

is less than the critical cavitation number σ_{crit}, which depends on the geometry and the Reynolds number. In Eq.(8.4), p_∞ is the absolute pressure in the free stream and p_v is the vapor pressure of the liquid.

The drag coefficient of a body that experiences cavitation is given by

$$C_D(\sigma) = C_D(0)(1 + \sigma) \tag{8.5}$$

where $C_D(0)$ is given in Table 8.3 for several bodies for $\text{Re} \cong 10^5$.

The *hydrofoil*, an airfoil-type shape that is used to lift a vessel above the water surface, invariably cannot operate without cavitation. The area and Reynolds number are based on the chord length. The drag and lift coefficients along with the critical cavitation numbers are presented in Table 8.4.

Table 8.3 Drag Coefficients for Zero Cavitation Numbers at Re $\cong 10^5$

Geometry	Angle	$C_D(0)$
Sphere		0.30
Disk (circular)		0.8
Cylinder (circular)		0.50
Flat plate (rectangular)		0.88
Two-dimensional wedge	120	0.74
	90	0.64
	60	0.49
	30	0.28
Cone (axisymmetric)	120	0.64
	90	0.52
	60	0.38
	30	0.20

Table 8.4 Drag and Lift Coefficients and Critical Cavitation Numbers for Hydrofoils for $10^5 < $ Re $< 10^6$

Angle (degrees)	Lift Coefficient	Drag Coefficient	Critical Cavitation Number
−2	0.2	0.014	0.5
0	0.4	0.014	0.6
2	0.6	0.015	0.7
4	0.8	0.018	0.8
6	0.95	0.022	1.2
8	1.10	0.03	1.8
10	1.22	0.04	2.5

EXAMPLE 8.3

A 2-m-long hydrofoil with chord length of 40 cm operates at 30 cm below the water's surface with an angle of attack of 6°. For a speed of 16 m/s determine the drag and lift and decide if cavitation exists on the hydrofoil.

Solution

The pressure p_∞ must be absolute. It is

$$p_\infty = \gamma h + p_{\text{atm}} = 9800 \times 0.3 + 100\,000 = 102\,900 \text{ Pa abs}$$

Assuming the water temperature is about 15°C, the vapor pressure is 1600 Pa (see Table C.1) so that the cavitation number is

$$\sigma = \frac{p_\infty - p_v}{\frac{1}{2}\rho V^2} = \frac{102\,900 - 1705}{0.5 \times 1000 \times 16^2} = 0.79$$

This is less than the critical cavitation number of 1.2 given in Table 8.4 so cavitation is present. Note: we could have used $p_v = 0$, as is often done, with sufficient accuracy. The drag and lift are

$$F_D = \frac{1}{2}\rho V^2 A C_D = \frac{1}{2} \times 1000 \times 16^2 \times 2 \times 0.4 \times 0.022 = 2250 \text{ N}$$

$$F_L = \frac{1}{2}\rho V^2 A C_L = \frac{1}{2} \times 1000 \times 16^2 \times 2 \times 0.4 \times 0.95 = 97\,300 \text{ N}$$

8.3 Flow Around Airfoils

Airfoils are streamlined so that separation does not occur. Airfoils designed to operate at subsonic speeds are rounded at the leading edge, whereas those designed for supersonic speeds may have sharp leading edges. The drag on an airfoil is primarily due to the shear stress that acts on the surface. The boundary layer, in which all the shear stresses are confined, that develops on an airfoil is very thin (see in Fig. 8.6) and can be ignored when solving for the inviscid flow surrounding the airfoil. The pressure distribution that is determined from the inviscid flow solution is influenced very little by the presence of the boundary layer. Consequently, the lift is estimated on an airfoil by ignoring the boundary layer and integrating the pressure distribution of the inviscid flow. The inviscid flow solution also provides the velocity at the outer edge of the thin boundary layer, a boundary condition needed when solving

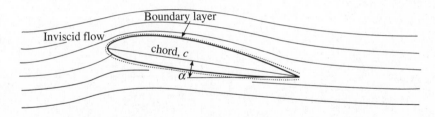

Figure 8.6   Flow around an airfoil at an angle of attack α.

the boundary-layer equations; the solution of the boundary-layer equations on a flat plate will be presented in Sec. 8.5.

The lift and drag on airfoils will not be calculated from the flow conditions but from graphical values of the lift and drag coefficients. These are displayed in Fig. 8.7 for a conventional airfoil with Re $\cong 9 \times 10^6$. The lift and drag coefficients are defined as

$$C_L = \frac{F_L}{\frac{1}{2}\rho c L V^2} \qquad C_D = \frac{F_D}{\frac{1}{2}\rho c L V^2} \tag{8.6}$$

Conventional airfoils are not symmetric and are designed to have positive lift at zero angle of attack, as shown in Fig. 8.7. The lift is directly proportional to the angle of attack until just before stall is encountered. The drag coefficient is also directly proportional to the angle of attack up to about 5°. The cruise condition is at an angle of attack of about 2°, where the drag is a minimum at $C_L = 0.3$ as noted. Mainly, the wings supply the lift on an aircraft. But an effective length is the tip-to-tip distance, the *wingspan,* since the fuselage also supplies lift.

The drag coefficient is essentially constant up to a Mach number of about 0.75. It then increases by over a factor of 10 until a Mach number of one is reached at which point it begins to slowly decrease. So, cruise Mach numbers between 0.75 and 1.5 are avoided to stay away from the high drag coefficients. Swept-back airfoils are used since it is the normal component of velocity that is used when calculating the Mach number; that allows a higher plane velocity before the larger drag coefficients are encountered.

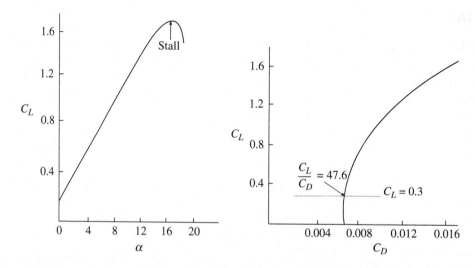

Figure 8.7   Lift and drag coefficients for a conventional airfoil at Re $\cong 9 \times 10^6$.

Slotted flaps are also used to provide larger lift coefficients during takeoff and landing. Air flows from the high-pressure region on the bottom of the airfoil through a slot to energize the slow-moving air in the boundary layer on the top side of the airfoil thereby reducing the tendency to separate and stall. The lift coefficient can reach 2.5 with a single-slotted flap and 3.2 with two slots.

EXAMPLE 8.4

Determine the takeoff speed for an aircraft that weighs $15\,000$ N including its cargo if its wingspan is 15 m with a 2-m chord. Assume an angle of attack of $8°$ at takeoff.

Solution

Assume a conventional airfoil and use the lift coefficient of Fig. 8.7 of about 0.95. The velocity is found from the equation for the lift coefficient:

$$C_L = \frac{F_L}{\frac{1}{2}\rho c L V^2} \qquad 0.95 = \frac{15\,000}{\frac{1}{2}\times 1.2 \times 2 \times 15 \times V^2} \qquad \therefore V = 30 \text{ m/s}$$

The answer is rounded off to two significant digits, since the lift coefficient of 0.95 is approximated from the figure.

8.4 Potential Flow

BASICS

When a body is moving in an otherwise stationary fluid, there is no vorticity present in the undisturbed fluid. To create a steady flow, a uniform flow with the body's velocity is superimposed on the flow field so that the vorticity-free flow moves by the stationary body, as in a wind tunnel. For high Reynolds-number flows, the viscous effects are concentrated in the boundary layer and the wake (the wake includes the separated region). For a streamlined body and over the front part of a blunt body, the flow outside the boundary layer is free of viscous effects so it is an inviscid flow. The solution of the inviscid flow problem provides the velocity field and the pressure field. The pressure is not significantly influenced by the boundary layer so the pressure integrated over the body's surface will provide the lift. The velocity at the surface of the body[2] from the inviscid flow solution will be the velocity at the

---

[2]If there are no viscous effects, the fluid does not stick to a boundary but is allowed to slip.

outer edge of the thin boundary layer needed in the boundary-layer solution (to be presented in Sec. 8.5). So, before the boundary layer can be analyzed on a body, the inviscid flow must be known.

A *potential flow* (or irrotational flow), is one in which the velocity field can be expressed as the gradient of a scalar function, that is,

$$\mathbf{V} = \nabla \phi \tag{8.7}$$

where ϕ is the *velocity potential*. For a potential flow, the vorticity is zero:

$$\omega = \nabla \times \mathbf{V} = \mathbf{0} \tag{8.8}$$

This can be shown to be true by expanding in rectangular coordinates and using Eq. (8.7).

To understand why vorticity cannot exist in regions of an irrotational flow, consider the effect of the three types of forces that can act on a cubic fluid element: the pressure and body forces act through the center of the element, and consequently, cannot impart a rotary motion to the element. It is only the viscous shear forces that are able to give rotary motion to fluid particles. Hence, if the viscous effects are nonexistent, vorticity cannot be introduced into an otherwise potential flow.

If the velocity is given by Eq. (8.7), the continuity equation (5.8) for an incompressible flow provides

$$\nabla \cdot \nabla \phi = \nabla^2 \phi = 0 \tag{8.9}$$

which is the famous *Laplace equation*. In rectangular coordinates it is written as

$$\frac{\partial^2 \phi}{\partial x^2} + \frac{\partial^2 \phi}{\partial y^2} + \frac{\partial^2 \phi}{\partial z^2} = 0 \tag{8.10}$$

With the required boundary conditions, this equation could be solved. But, rather than attempting to solve the resulting boundary-value problem directly, we will restrict our interest to plane flows, identify several simple flows that satisfy Laplace equation, and then superimpose those simple flows to form more complex flows of interest. Since Laplace equation is linear, the superimposed flows will also satisfy Laplace equation.

First, however, we will define another scalar function that will be quite useful in our study. For the plane flows of interest, the *stream function* ψ, is defined by

$$u = \frac{\partial \psi}{\partial y} \qquad \text{and} \qquad v = -\frac{\partial \psi}{\partial x} \tag{8.11}$$

so that the continuity equation (5.8) with $\partial w/\partial z = 0$ (for a plane flow) is satisfied for all plane flows. The vorticity [see Eqs. (8.8) and (3.14)] then provides

$$\omega_z = \frac{\partial v}{\partial x} - \frac{\partial u}{\partial y} = -\frac{\partial^2 \psi}{\partial x^2} - \frac{\partial^2 \psi}{\partial y^2} = 0 \tag{8.12}$$

so that

$$\frac{\partial^2 \psi}{\partial x^2} + \frac{\partial^2 \psi}{\partial y^2} = 0 \tag{8.13}$$

The stream function also satisfies the Laplace equation. So, from the above equations we have

$$u = \frac{\partial \phi}{\partial x} = \frac{\partial \psi}{\partial y} \quad \text{and} \quad v = \frac{\partial \phi}{\partial y} = -\frac{\partial \psi}{\partial x} \tag{8.14}$$

The equations between ϕ and ψ in Eq. (8.14) form the *Cauchy-Riemann equations* and ϕ and ψ are referred to as *harmonic functions*. The function $\phi + i\psi$ is the *complex velocity potential*. The mathematical theory of complex variables is thus applicable to this subset of fluid flows: steady, incompressible, inviscid plane flows.

Three items of interest contained in the above equations are:

- The stream function is constant along a streamline.
- The streamlines and lines of constant potential intersect at right angles.
- The difference of the stream functions between two streamlines is the flow rate q per unit depth between the two streamlines, i.e., $q = \psi_2 - \psi_1$.

EXAMPLE 8.5
Show that ψ is constant along a streamline.

Solution
A streamline is a line to which the velocity vector is tangent. This is expressed in vector form as $\mathbf{V} \times \mathbf{dr} = \mathbf{0}$, which, for a plane flow (no z variation), using $\mathbf{dr} = dx\,\mathbf{i} + dy\,\mathbf{j}$ takes the form $u\,dy - v\,dx = 0$. Using Eq. (8.11), this becomes

$$\frac{\partial \psi}{\partial y} dy + \frac{\partial \psi}{\partial x} dx = 0$$

This is the definition of $d\psi$ from calculus, thus $d\psi = 0$ along a streamline, or, in other words, ψ is constant along a streamline.

SEVERAL SIMPLE FLOWS

Several of the simple flows to be presented are much easier understood using polar (cylindrical) coordinates. The Laplace equation, the continuity equation, and the expressions for the velocity components for a plane flow (see Table 5.1) are

$$\nabla^2 \psi = \frac{1}{r}\frac{\partial}{\partial r}\left(r\frac{\partial \psi}{\partial r}\right) + \frac{1}{r^2}\frac{\partial^2 \psi}{\partial \theta^2} = 0 \tag{8.15}$$

$$\frac{1}{r}\frac{\partial}{\partial r}(rv_r) + \frac{1}{r}\frac{\partial v_\theta}{\partial \theta} = 0 \tag{8.16}$$

$$v_r = \frac{\partial \phi}{\partial r} = \frac{1}{r}\frac{\partial \psi}{\partial \theta} \quad \text{and} \quad v_\theta = \frac{1}{r}\frac{\partial \phi}{\partial \theta} = -\frac{\partial \psi}{\partial r} \tag{8.17}$$

where the expressions relating the velocity components to the stream function are selected so that the continuity equation is always satisfied. We now define four simple flows that satisfy the Laplace equation:

Uniform flow: $\quad \psi = U_\infty y \qquad \phi = U_\infty x$ $\tag{8.18}$

Line source: $\quad \psi = \frac{q}{2\pi}\theta \qquad \phi = \frac{q}{2\pi}\ln r$ $\tag{8.19}$

Vortex: $\quad \psi = \frac{\Gamma}{2\pi}\ln r \qquad \phi = \frac{\Gamma}{2\pi}\theta$ $\tag{8.20}$

Doublet: $\quad \psi = -\frac{\mu \sin\theta}{r} \qquad \phi = -\frac{\mu \cos\theta}{r}$ $\tag{8.21}$

These simple plane flows are shown in Fig. 8.8. If a y-component is desired for the uniform flow, an appropriate term is added. The *source strength q* in the line source is the flow rate per unit depth; adding a minus sign creates a sink. The *vortex strength* Γ is the *circulation* about the origin, defined as

$$\Gamma = \oint_L \mathbf{V} \cdot d\mathbf{s} \tag{8.22}$$

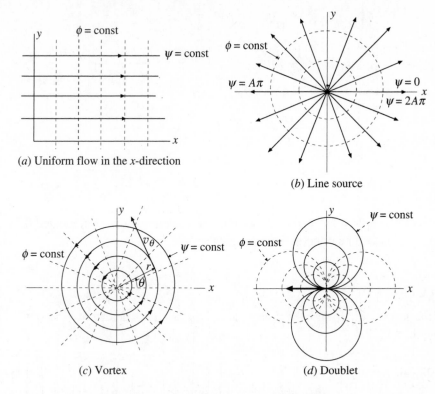

(a) Uniform flow in the x-direction

(b) Line source

(c) Vortex

(d) Doublet

Figure 8.8 Four simple plane potential flows.

where L is a closed curve, usually a circle, about the origin with clockwise being positive. The heavy arrow in the negative x-direction represents the *doublet strength* μ in Fig. 8.8d. (A doublet can be thought of as a source and a sink of equal strengths separated by a very small distance.)

The velocity components are used quite often for the four simple flows presented. They follow for both polar and rectangular coordinates:

$$\textit{Uniform flow:} \qquad \begin{aligned} u &= U_\infty & v &= 0 \\ v_r &= U_\infty \cos\theta & v_\theta &= -U_\infty \sin\theta \end{aligned} \qquad (8.23)$$

$$\textit{Line source:} \qquad \begin{aligned} v_r &= \frac{q}{2\pi r} & v_\theta &= 0 \\ u &= \frac{q}{2\pi}\frac{x}{x^2+y^2} & v &= \frac{q}{2\pi}\frac{y}{x^2+y^2} \end{aligned} \qquad (8.24)$$

$$\text{Vortex:} \quad \begin{array}{cc} v_r = 0 & v_\theta = -\dfrac{\Gamma}{2\pi r} \\ \\ u = -\dfrac{\Gamma}{2\pi} \dfrac{y}{x^2 + y^2} & v = \dfrac{\Gamma}{2\pi} \dfrac{x}{x^2 + y^2} \end{array} \quad (8.25)$$

$$\text{Doublet:} \quad \begin{array}{cc} v_r = -\dfrac{\mu \cos\theta}{r^2} & v_\theta = -\dfrac{\mu \sin\theta}{r^2} \\ \\ u = -\mu \dfrac{x^2 - y^2}{(x^2 + y^2)^2} & v = -\mu \dfrac{2xy}{(x^2 + y^2)^2} \end{array} \quad (8.26)$$

These four simple flows can be superimposed to create more complicated flows of interest. This will be done in the following section.

EXAMPLE 8.6
If the stream function of a flow is given as $\psi = A\theta$, determine the potential function ϕ.

Solution
We use Eq. (8.17) to relate the stream function to the potential function assuming polar coordinates because of the presence of θ:

$$\frac{\partial\phi}{\partial r} = \frac{1}{r}\frac{\partial\psi}{\partial\theta} = \frac{A}{r} \qquad \therefore \phi(r,\theta) = A\ln r + f(\theta)$$

Now, use the second equation of Eq. (8.17):

$$\frac{1}{r}\frac{\partial\phi}{\partial\theta} = \frac{1}{r}\frac{df}{d\theta} = -\frac{\partial\psi}{\partial r} = 0 \text{ implying that } \frac{df}{d\theta} = 0 \text{ so that } f = \text{const}$$

Since we are only interested in the derivatives of the potential functions needed to provide the velocity and pressure fields, we simply let the constant be zero and thus

$$\phi(r,\theta) = A\ln r$$

So, we see that the potential function can be found if the stream function is known. Conversely, the stream function can be found if the potential function is known.

SUPERIMPOSED FLOWS

The simple flows just defined can be superimposed to create complicated plane flows. Divide a surface, such as an airfoil, into a large number of segments and position sources or sinks (or doublets) at the center of each segment; in addition, add a uniform flow and a vortex. Then, adjust the various strengths so that the normal velocity component at each segment is zero and the rear stagnation point is located at the trailing edge. Obviously, a computer program would be used to create such a flow. We will not attempt it in this book but will demonstrate how flow around a circular cylinder can be created.

Superimpose the stream functions of a uniform flow and a doublet:

$$\psi(r,\theta) = U_\infty y - \frac{\mu \sin\theta}{r} \tag{8.27}$$

The velocity component v_r is (let $y = r\sin\theta$)

$$v_r = \frac{1}{r}\frac{\partial\psi}{\partial\theta} = U_\infty\cos\theta - \frac{\mu}{r^2}\cos\theta \tag{8.28}$$

A circular cylinder exists if there is a circle on which there is no radial velocity component, i.e., $v_r = 0$ at $r = r_c$. Set $v_r = 0$ in Eq. (8.28) and find

$$U_\infty\cos\theta - \frac{\mu}{r_c^2}\cos\theta = 0 \qquad \text{so that} \qquad r_c = \sqrt{\frac{\mu}{U_\infty}} \tag{8.29}$$

At this radius $v_r = 0$ for all θ and thus $r = r_c$ is a streamline and the result is flow around a cylinder. The stagnation points occur where the velocity is zero; if $r = r_c$ this means where $v_\theta = 0$, that is,

$$v_\theta = -\frac{\partial\psi}{\partial r}\bigg|_{r-r_c} = -U_\infty\sin\theta - \frac{\mu\sin\theta}{r_c^2} = 0 \qquad \therefore -2U_\infty\sin\theta = 0 \tag{8.30}$$

Thus, two stagnation points occur at $\theta = 0°$ and $180°$. The streamline pattern would appear as in the sketch of Fig. 8.9. The circular streamline represents the cylinder, which is typically a solid, and hence our interest is in the flow outside the circle. For a real flow, there would be a separated region on the rear of the cylinder but the flow over the front part (perhaps over the whole front half, depending on the Reynolds number) could be approximated by the potential flow shown in the figure. The velocity that exists outside the thin boundary layer that would be present on a real

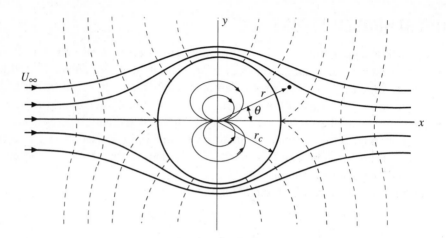

Figure 8.9   Potential flow around a circular cylinder. (The dashed lines are lines of constant ϕ.)

cylinder would be approximated as the velocity on the cylinder of the potential flow, that is, it would be given by

$$v_\theta = -2U_\infty \sin\theta \qquad (8.31)$$

The pressure that would exist on the cylinder's surface would be found by applying Bernoulli's equation between the stagnation point ($V = 0$) where the pressure is p_0 and some general point at r_c and θ:

$$p_c = p_0 - \rho\frac{v_\theta^2}{2} \qquad (8.32)$$

This pressure would approximate the actual pressure for high Reynolds-number flows up to separation. For low Reynolds-number flows, say below Re $\approx$ 50, viscous effects are not confined to a thin boundary layer so potential flow does not approximate the real flow.

    To create flow around a rotating cylinder, as in Fig. 8.10, add a vortex to the stream function of Eq. (8.27) [use the cylinder's radius of Eq. (8.29)]:

$$\psi(r,\theta) = U_\infty y - r_c^2 U_\infty \frac{\sin\theta}{r} + \frac{\Gamma}{2\pi}\ln r \qquad (8.33)$$

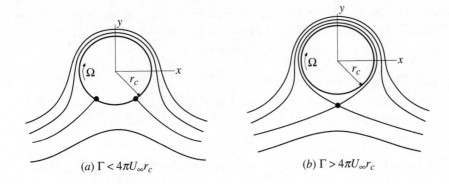

$(a)\ \Gamma < 4\pi U_\infty r_c$ $(b)\ \Gamma > 4\pi U_\infty r_c$

Figure 8.10 Flow around a rotating cylinder.

The cylinder's radius remains unchanged since a vortex does not affect v_r. The stagnation points change, however, and are located by letting $v_\theta = 0$ on $r = r_c$:

$$v_\theta = -\frac{\partial \psi}{\partial r}\bigg|_{r=r_c} = -U_\infty \sin\theta - r_c^2 U_\infty \frac{\sin\theta}{r_c^2} - \frac{\Gamma}{2\pi r_c} = 0 \tag{8.34}$$

This locates the stagnation points at

$$\theta = \sin^{-1}\frac{-\Gamma}{4\pi r_c U_\infty} \tag{8.35}$$

If $\Gamma > 4\pi r_c U_\infty$, Eq. (8.35) is not valid (this would give $|\sin\theta| > 1$), so the stagnation point exists off the cylinder as shown in Fig. 8.10b. The radius is found by setting the velocity components equal to zero and the angle $\theta = 270°$.

The pressure on the surface of the rotating cylinder is found using Bernoulli's equation to be

$$p_c = p_0 - \rho\frac{v_c^2}{2} = p_0 - \rho\frac{U_\infty^2}{2}\left(2\sin\theta + \frac{\Gamma}{2\pi r_c U_\infty}\right)^2 \tag{8.36}$$

If $p_c\,dA$ is integrated around the surface of the cylinder, the component in the flow direction, the drag, would be zero and the component normal to the flow direction, the lift, would be

$$F_L = \int_0^{2\pi} p_c \sin\theta\, r_c d\theta = \rho U_\infty \Gamma \tag{8.37}$$

It turns out that this expression for the lift is applicable for all cylinders including the airfoil. It is known as the *Kutta-Joukowski theorem;* it is exact for potential flows and is an approximation for real flows.

EXAMPLE 8.7

A 20-cm-diameter cylinder rotates clockwise at 200 rpm in an atmospheric air stream flowing at 10 m/s. Locate any stagnation points and find the minimum pressure.

Solution

First, find the circulation. It is $\Gamma = \oint_L \mathbf{V} \cdot d\mathbf{s}$ which is the velocity $r_c\Omega$ multiplied by $2\pi r_c$, since $\mathbf{V}$ is in the direction of $d\mathbf{s}$ on the cylinder's surface:

$$\Gamma = 2\pi r_c^2 \Omega = 2\pi \times 0.1^2 \times (200 \times 2\pi/60) = 1.316 \text{ m}^2/\text{s}$$

This is less than $4\pi r_c U_\infty = 12.57 \text{ m}^2/\text{s}$, so the two stagnation points are on the cylinder at

$$\theta = \sin^{-1} \frac{-\Gamma}{4\pi r_c U_\infty} = \sin^{-1} \frac{-1.316}{4\pi \times 0.1 \times 10} = -6° \quad \text{and} \quad 186°$$

The minimum pressure exists at the very top of the cylinder (see Fig. 8.10), so apply Bernoulli's equation [see Eq. (8.36)] between the free stream and the point on the top where $\theta = 90°$:

$$p_c = \cancel{p_\infty} + \rho \frac{U_\infty^2}{2} - \rho \frac{U_\infty^2}{2} \left(2\sin\theta + \frac{\Gamma}{2\pi r_c U_\infty} \right)^2$$

$$= 0 + 1.2 \times \frac{10^2}{2} \left[1 - \left(2\sin 90° + \frac{1.316}{2\pi \times 0.1 \times 10} \right)^2 \right] = -233 \text{ Pa}$$

using $\rho = 1.2 \text{ kg/m}^3$ for atmospheric air. (If the temperature is not given, assume standard conditions.)

8.5 Boundary-Layer Flow

GENERAL INFORMATION

The observation that for a high Reynolds-number flow all the viscous effects can be confined to a thin layer of fluid near the surface gives rise to boundary-layer theory. Outside the boundary layer the fluid acts as an inviscid fluid. So, the potential flow theory of the previous section provides both the velocity at the

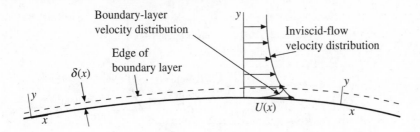

Figure 8.11   A boundary layer.

outer edge of the boundary layer and the pressure at the surface. In this section we will provide the integral and differential equations needed to solve for the velocity distribution in the boundary layer. But, those equations are quite difficult to solve for curved surfaces, so we will restrict our study to flow on a flat plate with zero pressure gradient.

The outer edge of a boundary layer cannot be observed, so we arbitrarily assign its thickness $\delta(x)$, as shown in Fig. 8.11, to be the locus of points where the velocity is 99 percent of the *free-stream velocity* $U(x)$ (the velocity at the surface from the inviscid flow solution). The pressure at the surface is not influenced by the presence of the thin boundary layer, so it is the pressure on the surface from the inviscid flow. Note that the xy-coordinate system is oriented so that the x-coordinate is along the surface; this is done for the boundary-layer equations and is possible because the boundary layer is so thin that curvature terms do not appear in the describing equations.

A boundary layer is laminar near the leading edge or near a stagnation point. It undergoes transition at x_T to a turbulent flow if there is sufficient length, as shown in Fig. 8.12. This transition occurs when the *critical Reynolds number* $U_\infty x_T/\nu = 5 \times 10^3$ on smooth, rigid flat plates in a zero pressure-gradient flow with low free-stream

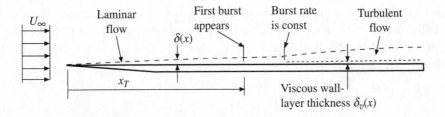

Figure 8.12   A boundary layer undergoing transition.

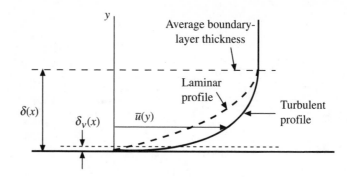

Figure 8.13   Laminar and turbulent boundary-layer profiles.

fluctuation intensity[3] and $U_\infty x_T/\nu = 3 \times 10^5$ for flow on rough flat plates or with high free-stream fluctuation intensity (intensity of at least 0.1). The transition region from laminar to turbulent flow is relatively short and is typically ignored so a turbulent flow is assumed to exist at the location of the first burst.

The turbulent boundary layer thickens more rapidly than a laminar boundary layer and contains significantly more momentum (if it has the same thickness), as observed from a sketch of the velocity profiles in Fig. 8.13. It also has a much greater slope at the wall resulting in a much larger wall shear stress. The instantaneous turbulent boundary layer varies randomly with time and position and can be 20 percent thicker or 60 percent thinner at any position at an instant in time or at any time at a given position. So, we usually sketch a time-average boundary-layer thickness. The *viscous wall layer* with thickness δ_ν in which turbulent bursts are thought to originate, is quite thin compared to the boundary-layer thickness, as shown.

It should be kept in mind that a turbulent boundary layer is very thin for most applications. On a flat plate with $U_\infty = 5$ m/s, the boundary layer would be about 7 cm thick after 4 m. If this were drawn to scale, the fact that the boundary layer is very thin would be quite apparent. Because the boundary layer is so thin and the velocity varies from zero at the wall to $U(x)$ at the edge of the boundary layer, it is possible to approximate the velocity profile in the boundary layer by assuming a parabolic or cubic profile for a laminar layer and a power-law profile for a turbulent layer. With the velocity profile assumed, the integral equations, which follow, give the quantities of interest.

---

[3]Fluctuation intensity is $\sqrt{\overline{u'^2}}/U_\infty$ [see Eq. (7.66)].

THE INTEGRAL EQUATIONS

An infinitesimal control volume of thickness dx is shown in Fig. 8.14 with mass fluxes in (b) and momentum fluxes in (d). The continuity equation provides the mass flux $\dot{m}_{\text{top}}$ that crosses into the control volume through the top; it is

$$\dot{m}_{\text{top}} = \dot{m}_{\text{out}} - \dot{m}_{\text{in}} = \frac{\partial}{\partial x}\left(\int_0^\delta \rho u\, dy\right) dx \qquad (8.38)$$

The x-component momentum equation (Newton's second law) is written as

$$\sum F_x = \dot{m}\dot{o}m_{\text{out}} - \dot{m}\dot{o}m_{\text{in}} - \dot{m}\dot{o}m_{\text{top}} \qquad (8.39)$$

which becomes

$$-\tau_0 dx - \delta dp = \frac{\partial}{\partial x}\left(\int_0^\delta \rho u^2 dy\right) dx - U(x)\frac{\partial}{\partial x}\left(\int_0^\delta \rho u\, dy\right) dx \qquad (8.40)$$

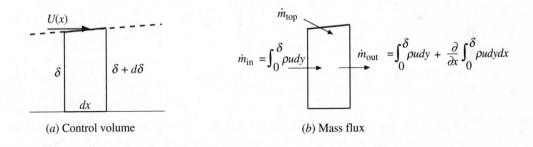

(a) Control volume

(b) Mass flux

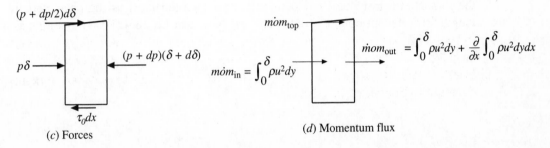

(c) Forces

(d) Momentum flux

Figure 8.14   The infinitesimal control volume for a boundary layer.

where we have neglected[4] $pd\delta$ and $dpd\delta$ since they are of smaller order than the remaining terms; we also used $\dot{m}om_{top} = U(x)\dot{m}_{top}$. Divide by $(-dx)$ and obtain the *von Karman integral equation*:

$$\tau_0 + \delta\frac{dp}{dx} = \rho U(x)\frac{d}{dx}\int_0^\delta u\,dy - \rho\frac{d}{dx}\int_0^\delta u^2\,dy \qquad (8.41)$$

Ordinary derivatives have been used since after the integration only a function of x remains (δ is a function of x). Also, the density ρ is assumed constant over the boundary layer.

For flow on a flat plate with zero pressure gradient, i.e., $U(x) = U_\infty$ and $\partial p/\partial x = 0$, Eq. (8.41) can be put in the simplified form

$$\tau_0 = \rho\frac{d}{dx}\int_0^\delta u(U_\infty - u)\,dy \qquad (8.42)$$

If a velocity profile $u(x, y)$ is assumed for a particular flow, Eq. (8.42) along with $\tau_0 = \mu\,\partial u/\partial y\big|_{y=0}$ allows both $\delta(x)$ and $\tau_0(x)$ to be determined.

Two additional lengths are used in the study of boundary layers. They are the *displacement thickness* δ_d and the *momentum thickness* θ defined by

$$\delta_d = \frac{1}{U}\int_0^\delta (U - u)\,dy \qquad (8.43)$$

$$\theta = \frac{1}{U^2}\int_0^\delta u(U - u)\,dy \qquad (8.44)$$

The displacement thickness is the distance the streamline outside the boundary layer is displaced because of the slower moving fluid inside the boundary layer. The momentum thickness is the thickness of a fluid layer with velocity U that possesses the momentum lost due to viscous effects; it is often used as the characteristic length for turbulent boundary-layer studies. Note that Eq. (8.42) can be written as

$$\tau_0 = \rho U_\infty^2\frac{d\theta}{dx} \qquad (8.45)$$

---

[4]$pd\delta$ is small since we assume δ to be small and $d\delta$ is then an order smaller.

LAMINAR AND TURBULENT BOUNDARY LAYERS

The boundary conditions that must be met for the velocity profile in a boundary layer on a flat plate with a zero pressure gradient are

$$
\begin{aligned}
u &= 0 && \text{at } y = 0 \\
u &= U_\infty && \text{at } y = \delta \\
\frac{\partial u}{\partial y} &= 0 && \text{at } y = \delta
\end{aligned}
\qquad (8.46)
$$

Laminar Boundary Layers

For a laminar boundary layer, we can either solve the x-component Navier-Stokes equation or we can assume a profile such as a parabola. Since the boundary layer is so thin, an assumed profile gives rather good results. Assume the parabolic profile

$$
\frac{u}{U_\infty} = A + By + Cy^2 \qquad (8.47)
$$

The above three boundary conditions require

$$
\begin{aligned}
0 &= A \\
1 &= A + B\delta + C\delta^2 \\
0 &= B + 2C\delta
\end{aligned}
\qquad (8.48)
$$

the solution of which is

$$
A = 0 \qquad B = \frac{2}{\delta} \qquad C = -\frac{1}{\delta^2} \qquad (8.49)
$$

This provides an estimate of the laminar-flow velocity profile

$$
\frac{u}{U_\infty} = 2\frac{y}{\delta} - \frac{y^2}{\delta^2} \qquad (8.50)
$$

Substitute this profile into the integral equation (8.42) and integrate:

$$
\tau_0 = \frac{d}{dx} \int_0^\delta \rho U_\infty^2 \left(\frac{2y}{\delta} - \frac{y^2}{\delta^2} \right)\left(1 - \frac{2y}{\delta} + \frac{y^2}{\delta^2} \right) dy = \frac{2}{15} \rho U_\infty^2 \frac{d\delta}{dx} \qquad (8.51)
$$

The wall shear stress is also given by

$$\tau_0 = \mu \left. \frac{\partial u}{\partial y} \right|_{y=0} = \mu U_\infty \frac{2}{\delta} \tag{8.52}$$

Equate the two expressions for τ_0 above to obtain

$$\delta d\delta = \frac{15\nu}{U} dx \tag{8.53}$$

Integrate the above with $\delta = 0$ at $x = 0$ and find the expression for $\delta(x)$:

$$\delta(x) = 5.48 \sqrt{\frac{\nu x}{U_\infty}} \quad \text{(approximate solution)} \tag{8.54}$$

This is about 10 percent higher than the more accurate solution of

$$\delta(x) = 5 \sqrt{\frac{\nu x}{U_\infty}} \quad \text{(accurate solution)} \tag{8.55}$$

found by solving the Navier-Stokes equation in the next section. The wall shear stress is found by substituting Eq. (8.54) into Eq. (8.52) and is

$$\tau_0(x) = 0.365 \rho U_\infty^2 \sqrt{\frac{\nu}{x U_\infty}} \tag{8.56}$$

The *local skin friction coefficient* c_f is often of interest and is

$$c_f(x) = \frac{\tau_0}{\frac{1}{2} \rho U_\infty^2} = 0.730 \sqrt{\frac{\nu}{x U_\infty}} \tag{8.57}$$

The *skin friction coefficient* C_f is a dimensionless drag force and is

$$C_f = \frac{F_D}{\frac{1}{2} \rho U_\infty^2 L} = \frac{\int_0^L \tau_0 dx}{\frac{1}{2} \rho U_\infty^2 L} = 1.46 \sqrt{\frac{\nu}{U_\infty L}} \tag{8.58}$$

The more accurate coefficients for τ_0, c_f, and C_f are 0.332, 0.664, and 1.33, so the assumption of a parabolic velocity profile for laminar boundary-layer flow has an error of about 10 percent.

Turbulent Boundary Layers

For a turbulent boundary layer we often assume a power-law velocity profile[5] as we did for flow in a pipe. It is

$$\frac{\bar{u}}{U_\infty} = \left(\frac{y}{\delta}\right)^{1/n} \qquad n = \begin{cases} 7 & \mathrm{Re}_x < 10^7 \\ 8 & 10^7 < \mathrm{Re}_x < 10^8 \\ 9 & 10^8 < \mathrm{Re}_x < 10^9 \end{cases} \qquad (8.59)$$

where $\mathrm{Re}_x = U_\infty x / v$. Substitute this velocity profile with $n = 7$ into Eq. (8.42) and integrate to obtain

$$\tau_0 = \frac{7}{72}\rho U_\infty^2 \frac{d\delta}{dx} \qquad (8.60)$$

The power-law velocity profile yields $\tau_0 = \mu \partial\bar{u}/\partial y = \infty$ at $y = 0$ so it cannot be used at the wall. A second expression for τ_0 is needed; we select the *Blasius formula*, given by

$$c_f = 0.046\left(\frac{v}{U_\infty\delta}\right)^{1/4} \qquad \text{giving} \qquad \tau_0 = 0.023\rho U_\infty^2\left(\frac{v}{U_\infty\delta}\right)^{1/4} \qquad (8.61)$$

Combine Eqs. (8.60) and (8.61) and find

$$\delta^{1/4}d\delta = 0.237\left(\frac{v}{U_\infty}\right)^{1/4}dx \qquad (8.62)$$

Assume a turbulent flow from the leading edge (the laminar portion is often quite short) and integrate from 0 to x:

$$\delta = 0.38x\left(\frac{v}{U_\infty x}\right)^{1/5} \qquad \mathrm{Re}_x < 10^7 \qquad (8.63)$$

---

[5]There are other more detailed and complicated methods for considering the turbulent boundary layer. They are all empirical since there are no analytical solutions of the turbulent boundary layer.

Substitute this into the Blasius formula and find the local skin friction coefficient to be

$$c_f = 0.059\left(\frac{v}{U_\infty x}\right)^{1/5} \qquad \text{Re}_x < 10^7 \tag{8.64}$$

The skin friction coefficient becomes

$$C_f = 0.073\left(\frac{v}{U_\infty L}\right)^{1/5} \qquad \text{Re}_x < 10^7 \tag{8.65}$$

The above formulas can actually be used up to $\text{Re} \cong 10^8$ without substantial error.

If there is a significant laminar part of the boundary layer, it should be included. If transition occurs at $\text{Re}_{\text{crit}} = 5 \times 10^5$, then the skin friction coefficient should be modified as

$$C_f = 0.073\left(\frac{v}{U_\infty L}\right)^{1/5} - 1700\frac{v}{U_\infty L} \qquad \text{Re}_x < 10^7 \tag{8.66}$$

For a rough plate, recall that $\text{Re}_{\text{crit}} = 3 \times 10^5$; the constant of 1700 in Eq. (8.66) should be replaced with 1060.

The displacement and momentum thicknesses can be evaluated using the power-law velocity profile to be

$$\delta_d = 0.048x\left(\frac{v}{U_\infty x}\right)^{1/5}$$
$$\qquad\qquad\qquad\qquad \text{Re}_x < 10^7 \tag{8.67}$$
$$\theta = 0.037x\left(\frac{v}{U_\infty x}\right)^{1/5}$$

There are additional quantities often used in the study of turbulent boundary layers. We will introduce two such quantities here. One is the *shear velocity* u_τ defined to be

$$u_\tau = \sqrt{\frac{\tau_0}{\rho}} \tag{8.68}$$

It is a fictitious velocity and often appears in turbulent boundary-layer relationships. The other is the thickness δ_v of the highly fluctuating viscous wall layer, displayed in Figs. 8.12 and 8.13. It is in this very thin layer that the turbulent bursts are thought to originate. It has been related to the shear velocity through experimental observations by

$$\delta_v = \frac{5v}{u_\tau} \qquad\qquad (8.69)$$

EXAMPLE 8.8
Atmospheric air at 20°C flows at 10 m/s over a smooth, rigid 2-m-wide, 4-m-long flat plate aligned with the flow. How long is the laminar portion of the boundary layer? Predict the drag force on the laminar portion on one side of the plate.

Solution
Assuming the air to free of high-intensity disturbances, use the critical Reynolds number to be 5×10^5, i.e.,

$$\frac{U_\infty x_T}{v} = 5 \times 10^5$$

so that

$$x_T = 5 \times 10^5 \times 1.51 \times 10^{-5}/10 = 0.755 \text{ m}$$

The drag force, using Eq. (8.58) and a coefficient of 1.33 rather than the 1.46 (the coefficient of 1.33 is more accurate as stated), is

$$F_D = \frac{1.33}{2} \rho U_\infty^2 \, Lw \sqrt{\frac{v}{U_\infty L}}$$

$$= 0.665 \times 1.2 \times 10^2 \times 0.755 \times 2 \times \sqrt{\frac{1.51 \times 10^{-5}}{10 \times 0.755}} = 0.17 \text{ N}$$

a rather small force.

EXAMPLE 8.9
Water at 20°C flows over a 2-m-long, 3-m-wide flat plate at 12 m/s. Estimate the shear velocity, the viscous wall-layer thickness, and the boundary-layer thickness at the end of the plate (assume a turbulent layer from the leading edge). Also, predict the drag force on one side of the plate.

Solution

The Reynolds number is $\text{Re} = U_\infty x / v = 12 \times 2 / 10^{-6} = 2.4 \times 10^7$. So, with $n = 7$ Eq. (8.64) provides

$$\tau_0 = \frac{0.059}{2} \rho U_\infty^2 \left(\frac{v}{U_\infty x} \right)^{1/5} = 0.0295 \times 1000 \times 12^2 \times \left(\frac{10^{-6}}{12 \times 2} \right)^{0.2} = 142 \text{ Pa}$$

The shear velocity is then

$$u_\tau = \sqrt{\frac{\tau_0}{\rho}} = \sqrt{\frac{142}{1000}} = 0.377 \text{ m/s}$$

The viscous wall-layer thickness is

$$\delta_v = \frac{5v}{u_\tau} = \frac{5 \times 10^{-6}}{0.377} = 1.33 \times 10^{-5} \text{ m}$$

The boundary-layer thickness is, assuming a turbulent layer from the leading edge,

$$\delta = 0.38 x \left(\frac{v}{U_\infty x} \right)^{1/5} = 0.38 \times 2 \times \left(\frac{10^{-6}}{12 \times 2} \right)^{0.2} = 0.0254 \text{ m}$$

The drag force on one side of the plate is

$$F_D = \frac{0.073}{2} \rho U_\infty^2 \, Lw \left(\frac{v}{U_\infty L} \right)^{1/5}$$

$$= 0.0365 \times 1000 \times 12^2 \times 2 \times 3 \times \left(\frac{10^{-6}}{12 \times 2} \right)^{0.2} = 1050 \text{ N}$$

LAMINAR BOUNDARY-LAYER DIFFERENTIAL EQUATIONS

The laminar flow solution given in the preceding section was an approximate solution. In this section we will present a more accurate solution using the x-component Navier-Stokes equation. It is, for horizontal plane flow (no z-variation),

$$u \frac{\partial u}{\partial x} + v \frac{\partial u}{\partial y} = -\frac{1}{\rho} \frac{\partial p}{\partial x} + v \left(\frac{\partial^2 u}{\partial x^2} + \frac{\partial^2 u}{\partial y^2} \right) \tag{8.70}$$

We can simplify this equation and actually obtain a solution. First, recall that the boundary layer is very thin so that there is no pressure variation normal to the boundary layer, i.e., the pressure depends on x only and it is the pressure at the wall from the potential flow solution. Since the pressure is considered known, the unknowns in Eq. (8.70) are u and v. The continuity equation

$$\frac{\partial u}{\partial x} + \frac{\partial v}{\partial y} = 0 \tag{8.71}$$

also relates u and v. So, we have two equations and two unknowns. Consider Figs. 8.12 and 8.13; u changes from zero to U_∞ over the very small distance δ, resulting in very large gradients in the y-direction, whereas u changes quite slowly in the x-direction (holding y fixed). Consequently, we conclude that

$$\frac{\partial^2 u}{\partial y^2} \gg \frac{\partial^2 u}{\partial x^2} \tag{8.72}$$

The differential equation (8.70) can then be written as

$$u\frac{\partial u}{\partial x} + v\frac{\partial u}{\partial y} = -\frac{1}{\rho}\frac{dp}{dx} + v\frac{\partial^2 u}{\partial y^2} \tag{8.73}$$

The two acceleration terms on the left side of the equation are retained, since v may be quite small but the gradient $\partial u/\partial y$ is quite large. Equation (8.73) is the *Prandtl boundary-layer equation*.

For flow on a flat plate with $dp/dx = 0$, and in terms of the stream function ψ (recall that $u = \partial\psi/\partial y$ and $v = -\partial\psi/\partial x$), Eq. (8.73) takes the form

$$\frac{\partial\psi}{\partial y}\frac{\partial^2\psi}{\partial x\partial y} - \frac{\partial\psi}{\partial x}\frac{\partial^2\psi}{\partial y^2} = v\frac{\partial^3\psi}{\partial y^3} \tag{8.74}$$

If we let (trial-and-error and experience were used to find this transformation)

$$\xi = x \qquad \text{and} \qquad \eta = y\sqrt{\frac{U_\infty}{vx}} \tag{8.75}$$

Eq. (8.74) becomes[6]

$$-\frac{1}{2\xi}\left(\frac{\partial\psi}{\partial\eta}\right)^2 + \frac{\partial\psi}{\partial\eta}\frac{\partial^2\psi}{\partial\xi\partial\eta} - \frac{\partial\psi}{\partial\xi}\frac{\partial^2\psi}{\partial\eta^2} = v\frac{\partial^3\psi}{\partial\eta^3}\sqrt{\frac{U_\infty}{v\xi}} \tag{8.76}$$

---

[6]Note that $\dfrac{\partial\psi}{\partial y} = \dfrac{\partial\psi}{\partial\eta}\dfrac{\partial\eta}{\partial y} + \dfrac{\partial\psi}{\partial\xi}\dfrac{\partial\xi}{\partial y} = \dfrac{\partial\psi}{\partial\eta}\sqrt{\dfrac{U_\infty}{vx}}$

This equation appears more formidable than Eq. (8.74), but if we let

$$\psi(\xi, \eta) = \sqrt{U_\infty \nu \xi}\, F(\eta) \tag{8.77}$$

and substitute this into Eq. (8.76), there results

$$F\frac{d^2 F}{d\eta^2} + 2\frac{d^3 F}{d\eta^3} = 0 \tag{8.78}$$

This ordinary differential equation can be solved numerically with the appropriate boundary conditions. They are

$$F = F' = 0 \text{ at } \eta = 0 \qquad \text{and} \qquad F' = 1 \text{ at large } \eta \tag{8.79}$$

which result from the velocity components

$$u = \frac{\partial \psi}{\partial y} = U_\infty F'(\eta)$$

$$v = -\frac{\partial \psi}{\partial x} = \frac{1}{2}\sqrt{\frac{\nu U_\infty}{x}}(\eta F' - F) \tag{8.80}$$

The numerical solution to the boundary-value problem is presented in Table 8.5. The last two columns allow the calculation of v and τ_0, respectively. We defined the

Table 8.5   The Laminar Boundary-Layer Solution with $dp/dx = 0$

$\eta = y\sqrt{\dfrac{U_\infty}{\nu x}}$	F	$F' = u/U_\infty$	$\frac{1}{2}(\eta F' - F)$	F''
0	0	0	0	0.3321
1	0.1656	0.3298	0.0821	0.3230
2	0.6500	0.6298	0.3005	0.2668
3	1.397	0.8461	0.5708	0.1614
4	2.306	0.9555	0.7581	0.0642
5	3.283	0.9916	0.8379	0.0159
6	4.280	0.9990	0.8572	0.0024
7	5.279	0.9999	0.8604	0.0002
8	6.279	1.000	0.8605	0.0000

boundary-layer thickness to be that thickness where $u = 0.99U_\infty$ and we observe that this occurs at $\eta = 5$. So, from this numerical solution,

$$\delta = 5\sqrt{\frac{vx}{U_\infty}} \tag{8.81}$$

Also,

$$\frac{\partial u}{\partial y} = \frac{\partial u}{\partial \eta}\frac{\partial \eta}{\partial y} = U_\infty F''\sqrt{\frac{U_\infty}{vx}} \tag{8.82}$$

so that the wall shear stress for this boundary layer with $dp/dx = 0$ is

$$\tau_0 = \mu\frac{\partial u}{\partial y}\bigg|_{y=0} = 0.332\rho U_\infty\sqrt{\frac{vU_\infty}{x}} \tag{8.83}$$

The friction coefficients are

$$c_f = 0.664\sqrt{\frac{v}{U_\infty x}} \qquad C_f = 1.33\sqrt{\frac{v}{U_\infty L}} \tag{8.84}$$

and the displacement and momentum thicknesses are (these require numerical integration)

$$\delta_d = 1.72\sqrt{\frac{vx}{U_\infty}} \qquad \theta = 0.644\sqrt{\frac{vx}{U_\infty}} \tag{8.85}$$

EXAMPLE 8.10
Air at 30°C flows over a 2-m-wide, 4-m-long flat plate with a velocity of 2 m/s and $dp/dx = 0$. At the end of the plate, estimate (a) the wall shear stress, (b) the maximum value of v in the boundary layer, and (c) the flow rate through the boundary layer. Assume laminar flow over the entire length.

Solution
The Reynolds number is $Re = U_\infty L/v = 2\times 4/1.6\times 10^{-5} = 5\times 10^5$ so laminar flow is reasonable.
(a) The wall shear stress (this requires F'' at the wall) at $x = 4$ m is

$$\tau_0 = 0.332\rho U_\infty\sqrt{\frac{vU_\infty}{x}} = 0.332\times 1.164\times 2\times\sqrt{\frac{1.6\times 10^{-5}\times 2}{4}} = 0.00219 \text{ Pa}$$

(b) The maximum value of v requires the use of $(\eta F' - F)$. Its maximum value occurs at the outer edge of the boundary layer and is 0.860. The maximum value of v is

$$v = \frac{1}{2}\sqrt{\frac{vU_\infty}{x}}(\eta F' - F) = \frac{1}{2} \times \sqrt{\frac{1.6 \times 10^{-5} \times 2}{4}} \times 0.860 = 0.0012 \text{ m/s}$$

Note the small value of v compared to $U_\infty = 2$ m/s.

(c) To find the flow rate through the boundary layer, integrate the $u(y)$ at $x = 4$ m:

$$Q = \int_0^\delta u \times 2dy = \int_0^5 U_\infty \frac{dF}{d\eta} \times 2 \times \sqrt{\frac{vx}{U_\infty}} d\eta$$

$$= 2 \times 2 \times \sqrt{\frac{1.6 \times 10^{-5} \times 4}{2}} \int_0^{3.283} dF = 0.0743 \text{ m}^3/\text{s}$$

Quiz No. 1

1. Estimate the drag coefficient for air at 10°C flowing around a 4.1-cm-diameter golf ball traveling at 35 m/s.

   (A) 0.25

   (B) 0.86

   (C) 0.94

   (D) 1.2

2. A fluid flows by a flat circular disk with velocity V normal to the disk with Re > 10^3. Estimate the drag coefficient if the pressure is assumed constant over the face of the disk. Assume the pressure is zero on the backside.

   (A) 1.1

   (B) 1.0

   (C) 0.9

   (D) 0.8

3. Atmospheric air at 20°C is flowing at 10 m/s normal to a 10-cm-wide, 20-cm-long rectangular plate. The drag force is nearest

   A) 2.95 N

   (B) 2.41 N

   (C) 1.76 N

   (D) 1.32 N

4. A 20-cm-diameter smooth sphere is rigged with a strain gage calibrated to measure the force on the sphere. Estimate the wind speed in 20°C air if the gage measures 0.5 N.

   (A) 21 m/s

   (B) 25 m/s

   (C) 29 m/s

   (D) 33 m/s

5. A 2.2-m-long hydrofoil with chord length of 50 cm operates 40 cm below the water's surface with an angle of attack of 4°. For a speed of 15 m/s, the lift is nearest

   (A) 75 kN

   (B) 81 kN

   (C) 99 kN

   (D) 132 kN

6. Estimate the takeoff speed for an aircraft with conventional airfoils if the aircraft with payload weighs 120 000 N and the effective wing area is 20 m² assuming a temperature of 30°C. An angle of attack at takeoff of 8° is desired.

   (A) 100 m/s

   (B) 95 m/s

   (C) 90 m/s

   (D) 85 m/s

7. Find the associated stream function if the potential function is $\phi = 10y$ in a potential flow.

   (A) $10x$

   (B) $10y$

   (C) $-10x$

   (D) $-10y$

8. A flow is represented by $\psi = 10\ln(x^2 + y^2)$. Find the pressure along the negative x-axis if atmospheric air is flowing and $p = 0$ at $x = -\infty$.

(A) $-120/x^2$

(B) $-160/x^2$

(C) $-200/x^2$

(D) $-240/x^2$

9. A uniform flow $\mathbf{V} = 10\mathbf{i}$ m/s is superimposed on a doublet with strength 40 m³/s. The velocity distribution $v_\theta(\theta)$ on the cylinder is

(A) $10\sin\theta$

(B) $20\sin\theta$

(C) $40\sin\theta$

(D) $80\sin\theta$

10. A turbulent boundary layer is studied in a zero pressure-gradient flow on a flat plate. Atmospheric air at 20°C flows over the plate at 10 m/s. How far from the leading edge can turbulence be expected if the free-stream fluctuation intensity is low?

(A) 60 cm

(B) 70 cm

(C) 80 cm

(D) 90 cm

11. Given $\tau_0 = 0.14\rho U_\infty^2 \, d\delta/dx$ where $\delta = 5\sqrt{vx/U_\infty}$ and $U_\infty = 1$ m/s. The drag force on one side of a 2-m-wide, 4-m-long plate, over which 20°C atmospheric air flows, is nearest

(A) 0.013 N

(B) 0.024 N

(C) 0.046 N

(D) 0.067 N

12. A laminar boundary layer of 20°C water moving at 0.8 m/s exists on one side of a 2-m-wide, 3-m-long flat plate. At $x = 3$ m, v_{max} is nearest

(A) 0.008 m/s

(B) 0.006 m/s

(C) 0.0004 m/s

(D) 0.0002 m/s

Quiz No. 2

1. Estimate the drag coefficient for air at 0°C moving past a 10-cm-diameter, 4-m-high pole at 2 m/s.

2. Atmospheric air at 20°C is flowing at 10 m/s normal to a 10-cm-diameter, 80-cm-long smooth cylinder with free ends. Calculate the drag force.

3. A 220-cm-square sign is impacted straight on by a 50 m/s 10°C wind. If the sign is held by a single 3-m-high post imbedded in concrete, what moment would exist at the base of the post?

4. A sensor is positioned downstream a short distance from a 4-cm-diameter cylinder in a 20°C atmospheric airflow. It senses vortex shedding at a frequency of 0.16 Hz. Estimate the airspeed.

5. A 2000-kg airplane is designed to carry a 4000-N payload when cruising near sea level. For a conventional airfoil with an effective wing area of 25 m², estimate the stall speed at elevation of 2000 m.

6. Find the associated potential function if the stream function is $\psi = 20xy$ in a potential flow.

7. A flow is represented by $\psi = 10\ln(x^2 + y^2)$ m²/s. Find the x-component of the acceleration at (−4, 0).

8. Locate any stagnation points in the flow represented by

$$\phi = 10r\cos\theta + 40\ln r \ \text{m}^2/\text{s}$$

9. A body is formed by the streamline that separates the source flow of strength $q = 10\pi$ m²/s from a uniform flow parallel to the x-axis of 10 m/s. Locate the positive y-intercept of the body formed by the streamline that separates the source flow from the uniform flow.

10. Assume a linear velocity profile in a laminar boundary layer on a flat plate with a zero pressure gradient. Find $\delta(x)$.

11. If the walls in a wind tunnel are parallel, the flow will accelerate due to the boundary layers on each of the walls. If a wind tunnel using 20°C atmospheric air is square, how should one of the walls be displaced outward for a zero pressure gradient to exist if $U = 10$ m/s? Assume turbulent flow.

12. Atmospheric air at 20°C flows over a 3-m-long and 2-m-wide flat plate at 16 m/s. Assume a turbulent flow from the leading edge and calculate drag force on one side of the plate.

13. A long cigar-shaped dirigible is proposed to take rich people on cruises. It is proposed to be 1000 m long and 150 m in diameter. How much horsepower is needed to move the dirigible through sea-level air at 12 m/s if the drag on the front and rear is neglected. Use the drag coefficient of Eq. (8.67).

CHAPTER 9

Compressible Flows

Compressible flows occur when the density changes are significant between two points on a streamline. Not all gas flows are compressible flows, only those that have significant density changes. Flow around automobiles, in hurricanes, around aircraft during landing and takeoff, and around buildings are a few examples of incompressible flows in which the density of the air does not change more than 3 percent between points of interest and are consequently treated as incompressible flows. There are, however, many examples of gas flows in which the density does change significantly; they include airflow around aircraft that fly faster than a Mach number [see Eq. (3.18)] of 0.3 (about 100 m/s), through compressors, jet engines, and tornados. Not considered in this chapter are compressible effects in liquid flows that give rise to water hammer and underwater compression waves from blasts.

9.1 Basics

Only compressible flow problems that can be solved using the integral equations will be considered in this chapter. The simplest of these is uniform flow in a conduit. Recall that the continuity equation, the momentum equation, and the energy equation (no work or heat transfer) are, respectively,

$$\dot{m} = \rho_1 A_1 V_1 = \rho_2 A_2 V_2 \tag{9.1}$$

$$\sum \mathbf{F} = \dot{m}(\mathbf{V}_2 - \mathbf{V}_1) \tag{9.2}$$

$$0 = \frac{V_2^2 - V_1^2}{2} + h_2 - h_1 \tag{9.3}$$

where the enthalpy $h = \tilde{u} + p/\rho$ is used. If the gas can be approximated as an ideal gas, then the energy equation takes either of the following two forms:

$$0 = \frac{V_2^2 - V_1^2}{2} + c_p(T_2 - T_1) \tag{9.4}$$

$$0 = \frac{V_2^2 - V_1^2}{2} + \frac{k}{k-1}\left(\frac{p_2}{\rho_2} - \frac{p_1}{\rho_1}\right) \tag{9.5}$$

where we have used the thermodynamic relations

$$\Delta h = c_p \Delta T \qquad c_p = c_v + R \qquad k = \frac{c_p}{c_v} \tag{9.6}$$

The ideal-gas law will also be used; the form most used is

$$p = \rho R T \tag{9.7}$$

We may also determine the entropy change or assume an isentropic process ($\Delta s = 0$). Then, one of the following equations may be used:

$$\Delta s = c_p \ln \frac{T_2}{T_1} - R \ln \frac{p_2}{p_1} \tag{9.8}$$

$$\frac{T_2}{T_1} = \left(\frac{p_2}{p_1}\right)^{(k-1)/k} \qquad \frac{p_2}{p_1} = \left(\frac{\rho_2}{\rho_1}\right)^k \qquad (\Delta s = 0) \tag{9.9}$$

Recall that the temperatures and pressures must always be absolute quantities when using several of the above relations. It is always safe to use absolute temperature and pressure when solving problems involving a compressible flow.

9.2 Speed of Sound

A pressure wave with small amplitude is called a *sound wave* and it travels through a gas with the *speed of sound*, denoted by c. Consider the small-amplitude wave shown in Fig. 9.1 traveling through a conduit. In Fig. 9.2a it is moving so that a stationary observer sees an unsteady motion; in Fig. 9.2b the observer moves with the wave so that the wave is stationary and a steady flow is observed; Fig. 9.2c shows the control volume surrounding the wave. The wave is assumed to create a small differential change in the pressure p, the temperature T, the density ρ, and the velocity V in the gas. The continuity equation applied to the control volume provides

$$\rho A c = (\rho + d\rho)A(c + dV) \tag{9.10}$$

which simplifies to, neglecting the higher-order term $d\rho\,dv$,

$$\rho dV = -c d\rho \tag{9.11}$$

The momentum equation in the streamwise direction is written as

$$pA - (p + dp)A = \rho A c(c + dV - c) \tag{9.12}$$

which simplifies to

$$dp = -\rho c dV \tag{9.13}$$

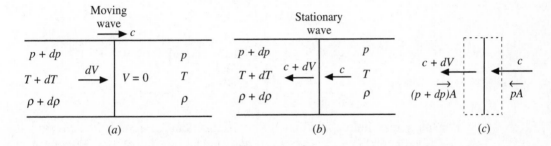

Figure 9.1 (*a*) A sound wave moving through a gas; (*b*) the gas moving through the wave; and (*c*) the control volume enclosing the wave of (*b*).

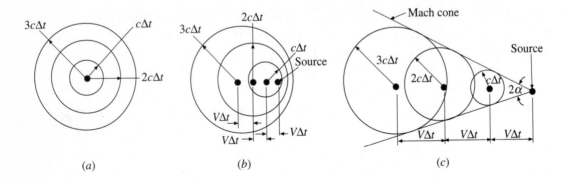

Figure 9.2   The propagation of sound waves from a source: (a) a stationary source; (b) a moving source with M < 1; (c) a moving source with M > 1.

Combining the continuity and momentum equations results in

$$c = \sqrt{\frac{dp}{d\rho}} \qquad (9.14)$$

for the small-amplitude sound waves.

The lower-frequency (less than 18 000 Hz) sound waves travel isentropically so that $p/\rho^k = \text{const}$ which, when differentiated, gives

$$\frac{dp}{d\rho} = k\frac{p}{\rho} \qquad (9.15)$$

The speed of sound for such waves is then

$$c = \sqrt{\frac{kp}{\rho}} = \sqrt{kRT} \qquad (9.16)$$

High-frequency waves travel isothermally resulting in a speed of sound of

$$c = \sqrt{RT} \qquad (9.17)$$

For small-amplitude waves traveling through a liquid or a solid, the bulk modulus is used [see Eq. (1.17)]; it is equal to $\rho\,dp/d\rho$ and has a value of 2100 MPa for water at 20°C. This gives a value for c of about 1450 m/s for a small-amplitude wave moving through water.

The Mach number, introduced in Chap. 3, is used for disturbances moving in a gas. It is

$$M = \frac{V}{c} \qquad (9.18)$$

If M < 1, the flow is *subsonic*, and if M > 1 the flow is *supersonic*. Consider the stationary source of disturbances displayed in Fig. 9.2a; the sound waves are shown after three time increments. In Fig. 9.2b the source is moving at a subsonic speed, which is less than the speed of sound so the source "announces" its approach to an observer to the right. In Fig. 9.2c the source moves at a supersonic speed, which is faster than the speed of the source, so an observer is unaware of the source's approach if the observer is in the *zone of silence*, which is outside the *Mach cone* shown. From the figure, the Mach cone has a *Mach angle* given by

$$\alpha = \sin^{-1}\frac{c}{V} = \sin^{-1}\frac{1}{M} \qquad (9.19)$$

The small-amplitude waves discussed above are referred to as *Mach waves*. They result from sources of sound, needle-nosed projectiles, and the sharp leading edge of supersonic airfoils. Large-amplitude waves, called *shock waves*, which emanate from the leading edge of blunt-nosed airfoils, also form zones of silence, but the angles are larger than those created by the Mach waves. Shock waves will be studied in a subsequent section.

EXAMPLE 9.1
An electronic device is situated on the top of a hill and hears a supersonic projectile that produces Mach waves after the projectile is 500 m past the device's position. If it is known that the projectile flies at 850 m/s, estimate how high it is above the device.

Solution
The Mach number is

$$M = \frac{V}{c} = \frac{850}{\sqrt{kRT}} = \frac{850}{\sqrt{1.4 \times 287 \times 288}} = 2.5$$

where a standard temperature of 288 K has been assumed since the temperature was not given. The Mach angle relationship allows us to write

$$\sin\alpha = \frac{1}{M} = \frac{h}{\sqrt{h^2 + 500^2}} = \frac{1}{2.5}$$

where h is the height above the device (refer to Fig. 9.2c). This equation can be solved for h to give

$$h = 218 \text{ m}$$

9.3 Isentropic Nozzle Flow

There are numerous applications where a steady, uniform, isentropic flow is a good approximation to the flow in conduits. These include the flow through a jet engine, through the nozzle of a rocket, from a broken gas line, and past the blades of a turbine. To model such situations, consider the control volume in the changing area of the conduit of Fig. 9.3. The continuity equation between two sections an infinitesimal distance dx apart is

$$\rho AV = (\rho + d\rho)(A + dA)(V + dV) \tag{9.20}$$

If only the first-order terms in a differential quantity are retained, continuity takes the form

$$\frac{dV}{V} + \frac{dA}{A} + \frac{d\rho}{\rho} = 0 \tag{9.21}$$

The energy equation (9.5) is

$$\frac{V^2}{2} + \frac{k}{k-1}\frac{p}{\rho} = \frac{(V+dV)^2}{2} + \frac{k}{k-1}\frac{p+dp}{\rho+d\rho} \tag{9.22}$$

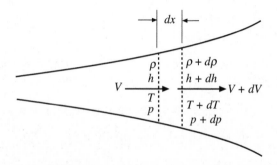

Figure 9.3 Steady, uniform, isentropic flow through a conduit.

This simplifies to, neglecting higher-order terms,

$$VdV + \frac{k}{k-1}\frac{\rho dp - p d\rho}{\rho^2} = 0 \qquad (9.23)$$

Assuming an isentropic flow, Eq. (9.15) allows the energy equation to take the form

$$VdV + k\frac{p}{\rho^2}d\rho = 0 \qquad (9.24)$$

Substitute from the continuity equation (9.21) to obtain

$$\frac{dV}{V}\left(\frac{\rho V^2}{kp} - 1\right) = \frac{dA}{A} \qquad (9.25)$$

or, in terms of the Mach number,

$$\frac{dV}{V}\left(M^2 - 1\right) = \frac{dA}{A} \qquad (9.26)$$

This equation applies to a steady, uniform, isentropic flow.

There are several observations that can be made from an analysis of Eq. (9.26). They are:

- For a subsonic flow in an expanding conduit ($M < 1$ and $dA > 0$), the flow is decelerating ($dV < 0$).
- For a subsonic flow in a converging conduit ($M < 1$ and $dA < 0$), the flow is accelerating ($dV > 0$).
- For a supersonic flow in an expanding conduit ($M > 1$ and $dA > 0$), the flow is accelerating ($dV > 0$).
- For a supersonic flow in a converging conduit ($M > 1$ and $dA < 0$), the flow is decelerating ($dV < 0$).
- At a throat where $dA = 0$, either $M = 1$ or $dV = 0$ (the flow could be accelerating through $M = 1$, or it may reach a velocity such that $dV = 0$).

Observe that a nozzle for a supersonic flow must increase in area in the flow direction, and a diffuser must decrease in area, opposite to a nozzle and diffuser for a subsonic flow. So, for a supersonic flow to develop from a reservoir where the velocity is zero, the subsonic flow must first accelerate through a converging area to a throat, followed by continued acceleration through an enlarging area.

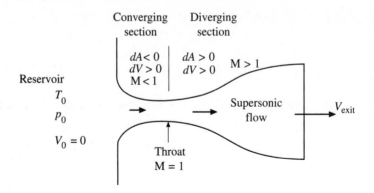

Figure 9.4 A supersonic nozzle.

The nozzles on a rocket designed to place satellites in orbit are constructed using such converging-diverging geometry, as shown in Fig. 9.4.

The energy and continuity equations can take on particularly helpful forms for the steady, uniform, isentropic flow through the nozzle of Fig. 9.4. Apply the energy equation (9.4) with $\dot{Q} = \dot{W}_S = 0$ between the reservoir and some location in the nozzle to obtain

$$c_p T_0 = \frac{V^2}{2} + c_p T \tag{9.27}$$

Any quantity with a zero subscript refers to a stagnation point where the velocity is zero, such as in the reservoir. Using several thermodynamic relations [Eqs. (9.6), (9.9), (9.16), and (9.18)], Eq. (9.27) can be put in the forms

$$\frac{T_0}{T} = 1 + \frac{k-1}{2} M^2$$

$$\frac{p_0}{p} = \left(1 + \frac{k-1}{2} M^2 \right)^{k/(k-1)} \tag{9.28}$$

$$\frac{\rho_0}{\rho} = \left(1 + \frac{k-1}{2} M^2 \right)^{1/(k-1)}$$

If the above equations are applied at the throat (the *critical area* signified by an asterisk (*) superscript, where M = 1), the energy equation takes the forms

$$\frac{T^*}{T_0} = \frac{2}{k+1} \qquad \frac{p^*}{p_0} = \left(\frac{2}{k+1} \right)^{k/(k-1)} \qquad \frac{\rho^*}{\rho_0} = \left(\frac{2}{k+1} \right)^{1/(k-1)} \tag{9.29}$$

The critical area is often referenced even though a throat does not exist, as in Table D.1. For air with $k = 1.4$, the equations above provide

$$T^* = 0.8333T_0 \qquad\qquad p^* = 0.5283p_0 \qquad\qquad \rho^* = 0.6340\rho_0 \qquad (9.30)$$

The mass flux through the nozzle is of interest and is given by

$$\dot{m} = \rho AV = \frac{p}{RT} \times A \times \mathrm{M}\sqrt{kRT} = p\sqrt{\frac{k}{RT}}A\mathrm{M} \qquad (9.31)$$

With the use of Eq. (9.28), the mass flux, after applying some algebra, can be expressed as

$$\dot{m} = p_0 \mathrm{M}A\sqrt{\frac{k}{RT_0}}\left(1 + \frac{k-1}{2}\mathrm{M}^2\right)^{(k+1)/2(1-k)} \qquad (9.32)$$

If the critical area is selected where $\mathrm{M} = 1$, this takes the form

$$\dot{m} = p_0 A^*\sqrt{\frac{k}{RT_0}}\left(1 + \frac{k-1}{2}\right)^{(k+1)/2(1-k)} \qquad (9.33)$$

which, when combined with Eq. (9.32), provides

$$\frac{A}{A^*} = \frac{1}{\mathrm{M}}\left[\frac{2 + (k-1)\mathrm{M}^2}{k+1}\right]^{(k+1)/2(1-k)} \qquad (9.34)$$

This ratio is included in the isentropic flow Table D.1 for air. The table can be used in place of the above equations.

Now we will discuss some features of the above equations. Consider a converging nozzle connecting a reservoir with a receiver, as shown in Fig. 9.5. If the reservoir pressure is held constant and the receiver pressure reduced, the Mach number at the exit of the nozzle will increase until $\mathrm{M}_e = 1$ is reached, indicated by the left curve in the figure. After $\mathrm{M}_e = 1$ is reached at the nozzle exit for $p_r = 0.5283p_0$, the condition of *choked flow* occurs and the velocity throughout the nozzle cannot change with further decreases in p_r. This is due to the fact that pressure changes downstream of the exit cannot travel upstream to cause changes in the flow conditions.

The right curve of Fig. 9.5*b* represents the case when the reservoir pressure is increased and the receiver pressure is held constant. When $\mathrm{M}_e = 1$, the condition of choked flow also occurs; but Eq. (9.33) indicates that the mass flux will continue to increase as p_0 is increased. This is the case when a gas line ruptures.

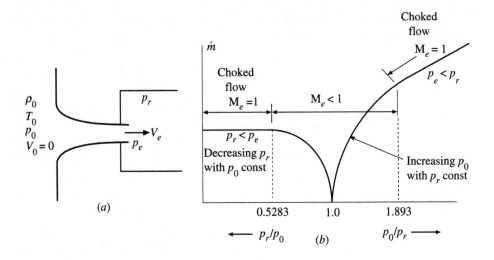

Figure 9.5 (*a*) A converging nozzle and (*b*) the pressure variation in the nozzle.

It is interesting that the exit pressure p_e is able to be greater than the receiver pressure p_r. Nature allows this by providing the streamlines of a gas the ability to make a sudden change of direction at the exit and expand to a much greater area resulting in a reduction of the pressure from p_e to p_r.

The case of a converging-diverging nozzle allows a supersonic flow to occur, providing the receiver pressure is sufficiently low. This is shown in Fig. 9.6 assuming a constant reservoir pressure with a decreasing receiver pressure. If the receiver pressure

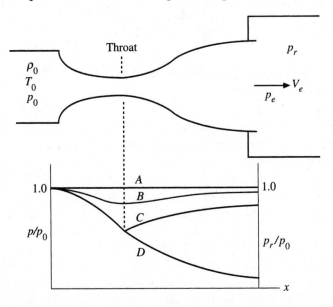

Figure 9.6 A converging-diverging nozzle with reservoir pressure fixed.

is equal to the reservoir pressure, no flow occurs, represented by curve A. If p_r is slightly less than p_0, the flow is subsonic throughout, with a minimum pressure at the throat, represented by curve B. As the pressure is reduced still further, a pressure is reached that results in M $= 1$ at the throat with subsonic flow throughout the remainder of the nozzle.

There is another receiver pressure substantially below that of curve C that also results in isentropic flow throughout the nozzle, represented by curve D; after the throat the flow is supersonic. Pressures in the receiver in between those of curve C and curve D result in non-isentropic flow (a shock wave occurs in the flow) and will be considered in the next section. If p_r is below that of curve D, the exit pressure p_e is greater than p_r. Once again, for receiver pressures below that of curve C, the mass flux remains constant since the conditions at the throat remain unchanged.

It may appear that the supersonic flow will tend to separate from the nozzle, but just the opposite is true. A supersonic flow can turn very sharp angles, as will be observed in Sec. 9.6, since nature provides expansion fans that do not exist in subsonic flows. To avoid separation in subsonic nozzles, the expansion angle should not exceed 10°. For larger angles, vanes are used so that the angle between the vanes does not exceed 10°.

EXAMPLE 9.2

Air flows from a reservoir maintained at 300 kPa absolute and 20°C into a receiver maintained at 200 kPa absolute by passing through a converging nozzle with an exit diameter of 4 cm. Calculate the mass flux through the nozzle. Use (a) the equations and (b) the isentropic flow table.

Solution

(a) The receiver pressure that would give M $= 1$ at the nozzle exit is

$$p_r = 0.5283 p_0 = 0.5283 \times 300 = 158.5 \text{ kPa abs}$$

The receiver pressure is greater than this so $\text{M}_e < 1$. The second equation of Eq. (9.28) can be put in the form

$$\frac{k-1}{2} \text{M}^2 = \left(\frac{p_0}{p}\right)^{(k-1)/k} - 1 \quad \text{or} \quad 0.2\text{M}^2 = \left(\frac{300}{200}\right)^{0.4/1.4} - 1$$

This gives M $= 0.784$. The mass flux is found from Eq. (9.32) to be

$$\dot{m} = p_0 \text{MA} \sqrt{\frac{k}{RT_0}} \left(1 + \frac{k-1}{2}\text{M}^2\right)^{(k+1)/2(1-k)}$$

$$= 300\,000 \times 0.784 \times \pi \times 0.02^2 \sqrt{\frac{1.4}{287 \times 293}} \left(1 + \frac{0.4}{2} \times 0.784^2\right)^{-2.4/0.8} = 0.852 \text{ kg/s}$$

For the units to be consistent, the pressure must be in Pa and R in J/kg·K.
(b) Now use Table D.1. For a pressure ratio of $p/p_0 = 200/300 = 0.6667$, the Mach number is found by interpolation to be

$$M_e = \frac{0.6821 - 0.6667}{0.6821 - 0.6560} \times (0.8 - 0.76) + 0.76 = 0.784$$

To find the mass flux the velocity must be known which requires the temperature since $V = M\sqrt{kRT}$. The temperature is interpolated (similar to the interpolation for the Mach number) from Table D.1 to be $T_e = 0.8906 \times 293 = 261$ K. The velocity and density are then

$$V = M\sqrt{kRT} = 0.784\sqrt{1.4 \times 287 \times 261} = 254 \text{ m/s}$$

$$\rho = \frac{p}{RT} = \frac{200}{0.287 \times 261} = 2.67 \text{ kg/m}^3$$

The mass flux is found to be

$$\dot{m} = \rho A V = 2.67 \times \pi \times 0.02^2 \times 254 = 0.852 \text{ kg/s}$$

9.4 Normal Shock Waves

Shock waves are large-amplitude waves that travel in a gas. They emanate from the wings of a supersonic aircraft, from a large explosion, from a jet engine, and ahead of the projectile in a gun barrel. They can be oblique waves or normal waves. First, we will consider the normal shock wave, as shown in Fig. 9.7. In this figure it is stationary so that a steady flow exists. If V_1 were superimposed to the left, the shock

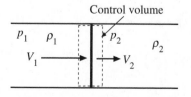

Figure 9.7 A stationary shock wave.

would be traveling in stagnant air with velocity V_1 and the *induced velocity* behind the shock wave would be $(V_1 - V_2)$. The shock wave is very thin, on the order of 10^{-4} mm, and in that short distance large pressure changes occur causing enormous energy dissipation. The continuity equation with $A_1 = A_2$ is

$$\rho_1 V_1 = \rho_2 V_2 \tag{9.35}$$

The energy equation with $\dot{Q} = \dot{W}_S = 0$ takes the form

$$\frac{V_2^2 - V_1^2}{2} + \frac{k}{k-1}\left(\frac{p_2}{\rho_2} - \frac{p_1}{\rho_1}\right) = 0 \tag{9.36}$$

The only forces in the momentum equation are pressure forces so

$$p_1 - p_2 = \rho_1 V_1 (V_2 - V_1) \tag{9.37}$$

where the areas have divided out since $A_1 = A_2$. Assuming that the three quantities $\rho_1, p_1,$ and V_1 before the shock wave are known, the above three equations allow us to solve for three unknowns $\rho_2, p_2,$ and V_2 since, for a given gas, k is known.

Rather than solve the above three equations simultaneously, we write them in terms of the Mach numbers M_1 and M_2, and put them in more convenient forms. First, the momentum equation (9.37), using Eq. (9.35) and $V^2 = M^2 pk/\rho$, can be written as

$$\frac{p_2}{p_1} = \frac{1 + kM_1^2}{1 + kM_2^2} \tag{9.38}$$

In like manner, the energy equation (9.36), with $p = \rho RT$ and $V^2 = M^2 kRT$, can be written as

$$\frac{T_2}{T_1} = \frac{1 + \dfrac{k-1}{2}M_1^2}{1 + \dfrac{k-1}{2}M_2^2} \tag{9.39}$$

The continuity equation (9.35) with $\rho = p/RT$ and $V = M\sqrt{kRT}$ becomes

$$\frac{p_2}{p_1}\frac{M_2}{M_1}\sqrt{\frac{T_1}{T_2}} = 1 \tag{9.40}$$

If the pressure and temperature ratios from Eqs (9.38) and (9.39) are substituted into Eq. (9.40) the downstream Mach number is related to the upstream Mach number by (the algebra to show this is complicated)

$$M_2^2 = \frac{M_1^2 + \dfrac{2}{k-1}}{\dfrac{2k}{k-1}M_1^2 - 1} \qquad (9.41)$$

This allows the momentum equation (9.38) to be written as

$$\frac{p_2}{p_1} = \frac{2k}{k+1}M_1^2 - \frac{k-1}{k+1} \qquad (9.42)$$

and the energy equation (9.39) as

$$\frac{T_2}{T_1} = \frac{\left(1 + \dfrac{k-1}{2}M_1^2\right)\left(\dfrac{2k}{k-1}M_1^2 - 1\right)}{\dfrac{(k+1)^2}{2(k-1)}M_1^2} \qquad (9.43)$$

For air, the preceding equations simplify to

$$M_2^2 = \frac{M_1^2 + 5}{7M_1^2 - 1} \qquad \frac{p_2}{p_1} = \frac{7M_1^2 - 1}{6} \qquad \frac{T_2}{T_1} = \frac{(M_1^2 + 5)(7M_1^2 - 1)}{36M_1^2} \qquad (9.44)$$

Several observations can be made from these three equations:

- If $M_1 = 1$, then $M_2 = 1$ and no shock wave exists.
- If $M_1 > 1$, then $M_2 < 1$ and supersonic flow is always converted to a subsonic flow when it passes through a normal shock wave.
- If $M_1 < 1$, then $M_2 > 1$ and a subsonic flow appears to be converted to a supersonic flow. This is impossible since it results in a positive production of entropy, a violation of the second law of thermodynamics.

Several normal shock-flow relations for air have been presented in Table D.2. The use of that table allows one to avoid using Eq. (9.44). In addition, the ratio p_{02}/p_{01} of the stagnation point pressures in front of and behind the shock wave are listed.

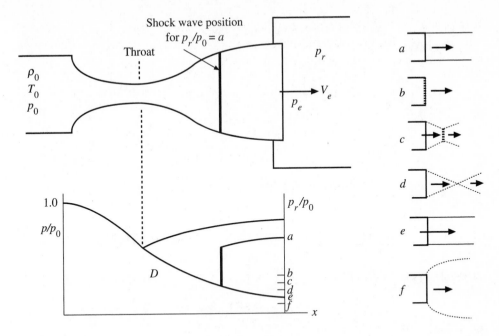

Figure 9.8 Flow with shock waves in a nozzle.

Return to the converging-diverging nozzle and focus attention on the flow below curve C of Fig. 9.6. If the receiver pressure decreases to $p_r/p_0 = a$ in Fig. 9.8, a normal shock wave would be positioned somewhere inside the nozzle as shown. If the receiver pressure decreased still further, there would be some ratio $p_r/p_0 = b$ that would position the shock wave at the exit plane of the nozzle. Pressure ratios c and d would result in oblique shock-wave patterns similar to those shown. Pressure ratio e is associated with isentropic flow throughout, and pressure ratio f would provide an exit pressure greater than the receiver pressure resulting in a billowing out, as shown, of the exiting flow, as seen on the rockets that propel satellites into space.

EXAMPLE 9.3

A normal shock wave travels at 600 m/s through stagnant 20°C air. Estimate the velocity induced behind the shock wave. (a) Use the equations and (b) use the normal shock-flow table D.2. Refer to Fig. 9.7.

Solution

Superimpose a velocity of 600 m/s so that the shock wave is stationary and $V_1 =$ 600 m/s, as displayed in Fig. 9.7. The upstream Mach number is

$$M_1 = \frac{V_1}{\sqrt{kRT}} = \frac{600}{\sqrt{1.4 \times 287 \times 293}} = 1.75$$

(a) Using the equations, the downstream Mach number and temperature are, respectively,

$$M_2 = \left(\frac{M_1^2 + 5}{7M_1^2 - 1}\right)^{1/2} = \left(\frac{1.75^2 + 5}{7 \times 1.75^2 - 1}\right)^{1/2} = 0.628$$

$$T_2 = \frac{T_1(M_1^2 + 5)(7M_1^2 - 1)}{36M_1^2} = \frac{293 \times (1.75^2 + 5)(7 \times 1.75^2 - 1)}{36 \times 1.75^2} = 438 \text{ K}$$

The velocity behind the shock wave is then

$$V_2 = M_2\sqrt{kRT_2} = 0.628 \times \sqrt{1.4 \times 287 \times 438} = 263 \text{ m/s}$$

If V_1 is superimposed to the left in Fig. 9.7, the induced velocity is

$$V_{induced} = V_1 - V_2 = 600 - 263 = 337 \text{ m/s}$$

which would act to the left, in the direction of the moving shock wave. (b) Table D.2 is interpolated at $M_1 = 1.75$ to find

$$M_2 = \frac{1.75 - 1.72}{1.76 - 1.72} \times (0.6257 - 0.6355) + 0.6355 = 0.6282$$

$$\frac{T_2}{T_1} = \frac{1.75 - 1.72}{1.76 - 1.72} \times (1.502 - 1.473) + 1.473 = 1.495 \qquad \therefore T_2 = 438 \text{ K}$$

The velocity V_2 is then

$$V_2 = M_2\sqrt{kRT_2} = 0.628 \times \sqrt{1.4 \times 287 \times 438} = 263 \text{ m/s}$$

and the induced velocity due to the shock wave is

$$V_{induced} = V_1 - V_2 = 600 - 263 = 337 \text{ m/s}$$

EXAMPLE 9.4
Air flows from a reservoir maintained at 20°C and 200 kPa absolute through a converging-diverging nozzle with a throat diameter of 6 cm and an exit diameter of 12 cm to a receiver. What receiver pressure is needed to locate a shock wave at a position where the diameter is 10 cm? Refer to Fig. 9.8.

Solution

Let's use the isentropic-flow table D.1 and the normal shock-flow table D.2. At the throat for this supersonic flow $M_t = 1$. The Mach number just before the shock wave is interpolated from Table D.1 where $A_1/A^* = 10^2/6^2 = 2.778$ to be

$$M_1 = 2.556$$

From Table D.2,

$$M_2 = 0.5078 \qquad \frac{p_{02}}{p_{01}} = 0.4778 \qquad \therefore p_{02} = 0.4778 \times 200 = 95.55 \text{ kPa}$$

since the stagnation pressure does not change in the isentropic flow before the shock wave so that $p_{01} = 200$ kPa. From just after the shock wave to the exit, isentropic flow again exists so that from Table D.1 at $M_2 = 0.5078$

$$\frac{A_2}{A^*} = 1.327$$

We have introduced an imaginary throat between the shock wave and the exit of the nozzle. The exit area A_e is introduced by

$$\frac{A_e}{A^*} = \frac{A_2}{A^*} \times \frac{A_e}{A_2} = 1.327 \times \frac{12^2}{10^2} = 1.911$$

Using Table D.1 at this area ratio (make sure the subsonic part of the table is used), we find

$$M_e = 0.3223 \qquad \text{and} \qquad \frac{p_e}{p_{0e}} = 0.9305 \qquad \therefore p_e = 0.9305 \times 95.55 = 88.9 \text{ kPa}$$

using $p_{0e} = p_{02}$ for the isentropic flow after the shock wave. The exit pressure is equal to the receiver pressure for this isentropic subsonic flow.

9.5 Oblique Shock Waves

Oblique shock waves form on the leading edge of a supersonic sharp-edged airfoil or in a corner, as shown in Fig. 9.9. A steady, uniform plane flow exists before and after the shock wave. The oblique shock waves also form on axisymmetric projectiles.

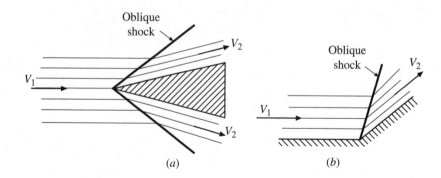

Figure 9.9   Oblique shock waves: (a) flow over a wedge and (b) flow in a corner.

The oblique shock wave turns the flow so that V_2 is parallel to the plane surface. Another variable, the angle through which the flow turns, is introduced but the additional tangential momentum equation allows a solution. Consider the control volume of Fig. 9.10 surrounding the oblique shock wave. The velocity vector $\mathbf{V}_1$ is assumed to be in the x-direction and the oblique shock wave turns the flow through the *wedge angle* or *deflection angle* θ so that $\mathbf{V}_2$ is parallel to the wall. The oblique shock wave makes an angle of β with $\mathbf{V}_1$. The components of the velocity vectors are shown normal and tangential to the oblique shock. The tangential components of the velocity vectors do not cause fluid to flow into or out of the control volume, so continuity provides

$$\rho_1 V_{1n} = \rho_2 V_{2n} \tag{9.45}$$

The pressure forces act normal to the control volume and produce no net force tangential to the oblique shock. This allows the tangential momentum equation to take the form

$$\dot{m}_1 V_{1t} = \dot{m}_2 V_{2t} \tag{9.46}$$

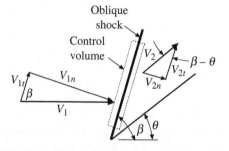

Figure 9.10   Oblique shock-wave control volume.

Continuity requires $\dot{m}_1 = \dot{m}_2$ so that

$$V_{1t} = V_{2t} \tag{9.47}$$

The momentum equation normal to the oblique shock is

$$p_1 - p_2 = \rho_2 V_{2n}^2 - \rho_1 V_{1n}^2 \tag{9.48}$$

The energy equation, using $V^2 = V_n^2 + V_t^2$, can be written in the form

$$\frac{V_{1n}^2}{2} + \frac{k-1}{k}\frac{p_1}{\rho_1} = \frac{V_{2n}^2}{2} + \frac{k-1}{k}\frac{p_2}{\rho_2} \tag{9.49}$$

since the tangential velocity terms cancel.

Observe that the tangential velocity components do not enter the three Eqs. (9.45), (9.48), and (9.49). They are the same three equations used to solve the normal shock-wave problem. So, the components V_{1n} and V_{2n} can be replaced with V_1 and V_2, respectively, of the normal shock-wave problem and a solution obtained. Table D.2 may also be used. We also replace M_{1n} and M_{2n} with M_1 and M_2 in the equations and table.

To often simplify a solution, we relate the oblique shock angle β to the deflection angle θ. This is done by using Eq. (9.45) to obtain

$$\frac{p_2}{\rho_1} = \frac{V_{1n}}{V_{2n}} = \frac{V_{1t}\tan\beta}{V_{2t}\tan(\beta-\theta)} = \frac{\tan\beta}{\tan(\beta-\theta)} \tag{9.50}$$

Using Eqs. (9.42) and (9.43) this density ratio can be written as

$$\frac{\rho_2}{\rho_1} = \frac{p_2 T_1}{p_1 T_2} = \frac{(k+1)M_{1n}^2}{(k-1)M_{1n}^2 + 2} \tag{9.51}$$

Using this density ratio in Eq. (9.50) allows us to write

$$\tan(\beta-\theta) = \frac{\tan\beta}{k+1}\left(k-1+\frac{2}{M_1^2\sin^2\beta}\right) \tag{9.52}$$

With this relationship the oblique shock angle β can be found for a given incoming Mach number and wedge angle θ. A plot of Eq. (9.52) is useful to avoid a trial-and-error solution. It is included as Fig. 9.11. Three observations can be made by studying the figure.

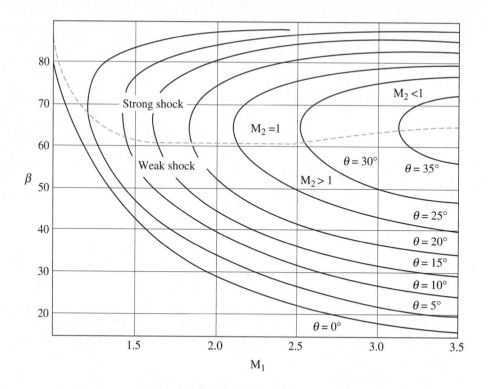

Figure 9.11 Oblique shock wave angle β related to wedge angle θ and Mach number M_1 for air.

- For given Mach number M_1 and wedge angle θ there are two possible oblique shock angles β. The larger one is the "strong" oblique shock wave and the smaller one is the "weak" oblique shock wave.

- For a given wedge angle θ there is a minimum Mach number for which there is only one oblique shock angle β.

- If the Mach number is less than the minimum for a particular θ, but greater than one, the shock wave is detached as shown in Fig. 9.12. Also, for a given M_1 there is a sufficiently large θ that will result in a detached shock wave.

The required pressure rise determines if a weak shock or a strong shock exists. The pressure rise is determined by flow conditions.

For a detached shock wave around a blunt body or a wedge, a normal shock wave exists on the stagnation streamline; the normal shock is followed by a strong oblique shock, then a weak oblique shock, and finally a Mach wave, as shown in Fig. 9.12. The shock wave is always detached on a blunt object.

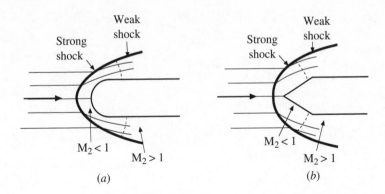

Figure 9.12   Detached shock waves around (a) a plane, blunt object and (b) a wedge.

EXAMPLE 9.5
Air at 30°C flows around a wedge with an included angle of 60° (see Fig. 9.9a).
An oblique shock emanates from the wedge at an angle of 50°. Determine the
approach velocity of the air. Also find M_2 and T_2.

Solution
From Fig. 9.11 the Mach number, at $\theta = 30°$ and $\beta = 50°$, is

$$M_1 = 3.1$$

The velocity is then

$$V_1 = M_1 \sqrt{kRT} = 3.1 \times \sqrt{1.4 \times 287 \times 303} = 1082 \text{ m/s}$$

If Eq. (9.52) were used for greater accuracy, we have

$$\tan(50° - 30°) = \frac{\tan 50°}{1.4 + 1}\left(1.4 - 1 + \frac{2}{M_1^2 \sin^2 50°}\right) \qquad \therefore M_1 = 3.20$$

The velocity would be $V_1 = 1117$ m/s.
   To find M_2, the approaching normal velocity and Mach number are

$$V_{1n} = V_1 \sin\beta = 1117 \sin 50° = 856 \text{ m/s} \qquad \therefore M_{1n} = \frac{856}{\sqrt{1.4 \times 287 \times 303}} = 2.453$$

From Table D.2 interpolation provides $M_{2n} = 0.5176$ so that

$$M_2 = \frac{M_{2n}}{\sin(50° - 30°)} = \frac{0.5176}{\sin 20°} = 1.513$$

The temperature behind the oblique shock is interpolated to be

$$T_2 = T_1 \times 2.092 = 303 \times 2.092 = 634 \text{ K}$$

9.6 Expansion Waves

Supersonic flow exits a nozzle (the pressure ratio f in Fig. 9.8), and billows out into a large exhaust plume. Also, supersonic flow does not separate from the wall of a nozzle that expands quite rapidly, as shown in Fig. 9.8. How is this accomplished? Consider the possibility that a single finite wave, such as an oblique shock, is able to turn the flow around the convex corner, as shown in Fig. 9.13a. From the tangential momentum equation, the tangential component of velocity must remain the same on both sides of the finite wave. For this to be true, $V_2 > V_1$. As before, this increase in velocity as the fluid flows through a finite wave requires an increase in entropy, a violation of the second law of thermodynamics, making a finite wave an impossibility.

A second possibility is to allow an infinite fan of Mach waves, called an *expansion fan*, emanating from the corner, as shown in Fig. 9.13b. This is an ideal isentropic process so the second law is not violated; such a process may be approached in a real application. Let's consider the single infinitesimal Mach wave displayed in Fig. 9.14, apply our fundamental laws, and then integrate around the corner. Since the tangential velocity components are equal, the velocity triangles yield

$$V_t = V \cos \mu = (V + dV)\cos(\mu + d\theta) \tag{9.53}$$

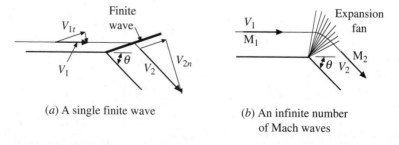

(a) A single finite wave　　　(b) An infinite number of Mach waves

Figure 9.13　Supersonic flow around a convex corner.

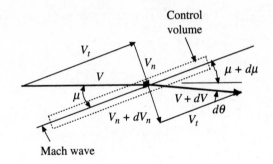

Figure 9.14   A Mach wave in an expansion fan.

This can be written as[1], neglecting higher-order terms,

$$V d\theta \sin \mu = \cos \mu \, dV \qquad (9.54)$$

Substitute $\sin \mu = 1/\mathrm{M}$ [see Eq. (9.19)] and $\cos \mu = \sqrt{\mathrm{M}^2 - 1}/\mathrm{M}$, to obtain

$$d\theta = \sqrt{\mathrm{M}^2 - 1} \frac{dV}{V} \qquad (9.55)$$

Differentiate the equation $V = \mathrm{M}\sqrt{kRT}$ and put in the form

$$\frac{dV}{V} = \frac{d\mathrm{M}}{\mathrm{M}} + \frac{1}{2}\frac{dT}{T} \qquad (9.56)$$

The energy equation $V^2/2 + kRT/(k-1) = \text{const}$ can also be differentiated to yield

$$\frac{dV}{V} + \frac{1}{(k-1)\mathrm{M}^2}\frac{dT}{T} = 0 \qquad (9.57)$$

Combine Eqs. (9.56) and (9.57) to obtain

$$\frac{dV}{V} = \frac{2}{2 + (k-1)\mathrm{M}^2}\frac{d\mathrm{M}}{\mathrm{M}} \qquad (9.58)$$

---

[1]Recall that $\cos(\mu + d\theta) = \cos \mu \cos d\theta - \sin \mu \sin d\theta = \cos \mu - d\theta \sin \mu$ since $\cos d\theta \approx 1$ and $\sin d\theta \approx d\theta$.

Substitute this into Eq. (9.55) to obtain a relationship between θ and M:

$$d\theta = \frac{2\sqrt{M^2-1}}{2+(k-1)M^2}\frac{dM}{M} \tag{9.59}$$

This is integrated from $\theta = 0$ and M = 1 to a general angle θ, called the *Prandtl-Meyer function,* and Mach number M (this would be M_2 in Fig. 9.13*b*) to find that

$$\theta = \left(\frac{k+1}{k-1}\right)^{1/2}\tan^{-1}\left[\frac{k-1}{k+1}(M^2-1)\right]^{1/2} - \tan^{-1}(M^2-1)^{1/2} \tag{9.60}$$

The solution to this relationship is presented for air in Table D.3 to avoid a trial-and-error solution for M given the angle θ. If the pressure or temperature is desired, the isentropic flow table can be used. The Mach waves that allow the gas to turn the corner are sometimes referred to as *expansion waves.*

Observe from Table D.3 that the expansion fan that turns the gas through the angle θ results in M = 1 before the fan to a supersonic flow after the fan. The gas speeds up as it turns the corner and it does not separate. A slower moving subsonic flow would separate from the corner and would slow down. If M = ∞ is substituted into Eq. (9.60), $\theta = 130.5°$, which is the maximum angle through which the flow could possibly turn. This shows that turning angles greater than 90° are possible, a rather surprising result.

EXAMPLE 9.6
Air at 150 kPa and 140°C flows at M = 2 and turns a convex corner of 30°. Estimate the Mach number, pressure, temperature, and velocity after the corner.

Solution
Table D.3 assumes the air is initially at M = 1. So, assume the flow originates from M = 1 and turns a corner to M_1 = 2 and then a second corner to M_2, as shown. From Table D.3, an angle of 26.4° is required to accelerate the flow from M = 1 to M = 2. Add another 30° to 26.4° and at θ = 56.4° we find that

$$M_2 = 3.37$$

Using the isentropic flow table D.1, the entries from the reservoir to state 1 and also to state 2 can be used to find

$$p_2 = p_1 \frac{p_0}{p_1} \frac{p_2}{p_0} = 150 \times \frac{1}{0.1278} \times 0.01580 = 18.54 \text{ kPa}$$

$$T_2 = T_1 \frac{T_0}{T_1} \frac{T_2}{T_0} = 413 \times \frac{1}{0.5556} \times 0.3058 = 227 \text{ K} \qquad \text{or} \qquad -46°\text{C}$$

The velocity after the corner is then

$$V_2 = M_2 \sqrt{kRT_2} = 3.37 \times \sqrt{1.4 \times 287 \times 227} = 1018 \text{ m/s}$$

Quiz No. 1

1. Two rocks are slammed together by a friend on one side of a lake. A listening device picks up the wave generated 0.45 s later. The distance across the lake is nearest

    (A) 450 m

    (B) 550 m

    (C) 650 m

    (D) 750 m

2. The Mach number for a projectile flying at 10 000 m at 200 m/s is

    (A) 0.62

    (B) 0.67

    (C) 0.69

    (D) 0.74

3. A supersonic aircraft passes 200 m overhead on a day when the temperature is 26°C. Estimate how far the aircraft is from you when you hear its sound if its Mach number is 1.68.

    (A) 245 m

    (B) 275 m

    (C) 315 m

    (D) 335 m

4. A converging nozzle with exit area of 10 cm² is attached to a reservoir maintained at 250 kPa absolute and 20°C. If the receiver pressure is maintained at 150 kPa absolute, the mass flux is nearest

   (A) 0.584 kg/s

   (B) 0.502 kg/s

   (C) 0.428 kg/s

   (D) 0.386 kg/s

5. Air flows through a converging-diverging nozzle attached from a reservoir maintained at 400 kPa absolute and 20°C to a receiver. If the throat and exit diameters are 10 and 24 cm, respectively, the receiver pressure that will just result in supersonic flow throughout is nearest

   (A) 6.8 kPa

   (B) 9.2 kPa

   (C) 16.4 kPa

   (D) 28.2 kPa

6. The temperature, pressure, and velocity before a normal shock wave in air are 18°C, 100 kPa absolute, and 600 m/s, respectively. The velocity after the shock wave is nearest

   (A) 212 m/s

   (B) 249 m/s

   (C) 262 m/s

   (D) 285 m/s

7. A large explosion occurs on the earth's surface producing a shock wave that travels radially outward. At a particular location the Mach number of the wave is 2.0. Determine the induced velocity behind the shock wave if $T_1 = 15°C$.

   (A) 502 m/s

   (B) 425 m/s

   (C) 368 m/s

   (D) 255 m/s

8. Air flows from a reservoir through a nozzle into a receiver. The reservoir is maintained at 400 kPa absolute and 20°C. The nozzle has a 10-cm-diameter throat and a 20-cm-diameter exit. The receiver pressure needed to locate a shock wave at the exit is nearest

    (A) 150 kPa

    (B) 140 kPa

    (C) 130 kPa

    (D) 120 kPa

9. A supersonic airflow changes direction 20° due to a sudden corner (see Fig. 9.9b). If $T_1 = 40$°C, $p_1 = 60$ kPa absolute, and $V_1 = 900$ m/s, calculate p_2 assuming a weak shock.

    (A) 164 kPa

    (B) 181 kPa

    (C) 192 kPa

    (D) 204 kPa

10. An airflow with a Mach number of 2.4 turns a convex corner of 40°. If the temperature and pressure are 5°C and 60 kPa absolute, respectively, the Mach number after the corner is nearest

    (A) 5

    (B) 3

    (C) 2

    (D) 1

Quiz No. 2

1. An underwater animal generates a signal that travels through water until it hits an object and then echoes back to the animal 0.46 s later. How far is the animal from the object?

2. A bolt of lightning lights up the sky and 1.5 s later you hear the thunder. How far did the lightning strike from your position?

3. A supersonic aircraft passes 200 m overhead on a day when the temperature is 26°C. Estimate how far the aircraft is from you when you hear its sound if its Mach number is 3.49.

4. A small-amplitude wave travels through the 15°C atmosphere creating a pressure rise of 5 Pa. Estimate the temperature rise across the wave and the induced velocity behind the wave.

5. A converging nozzle with exit area of 10 cm² is attached to a reservoir maintained at 250 kPa absolute and 20°C. Calculate the mass flux if the receiver pressure is maintained at 100 kPa absolute.

6. Air flows from a converging-diverging nozzle from a reservoir maintained at 400 kPa absolute and 20°C through a 12-cm-diameter throat. At what diameter in the diverging section will $M = 2$?

7. The temperature and pressure before a normal shock wave in air are 20°C and 400 kPa absolute, respectively. The Mach number after the shock wave is 0.5. Calculate the pressure and velocity after the shock.

8. Air flows from a reservoir maintained at 400 kPa absolute and 20°C out a nozzle with a 10-cm-diameter throat and a 20-cm-diameter exit into a receiver. Estimate the receiver pressure needed to locate a shock wave at a diameter of 16 cm.

9. A supersonic airflow changes direction 20° due to a sudden corner (see Fig. 9.9b). If $T_1 = 40°C$, $p_1 = 60$ kPa absolute, and $V_1 = 900$ m/s, calculate p_2 and V_2 assuming a strong shock.

10. An airflow with $M = 3.6$ is desired by turning a 20°C-supersonic flow with a Mach number of 1.8 around a convex corner. If the upstream pressure is 40 kPa absolute, what angle should the corner possess? What is the velocity after the corner?

APPENDIX A

Units and Conversions

Table A.1 English Units, SI Units, and Their Conversion Factors

Quantity	English Units	International System* SI	Conversion Factor
Length	inch	millimeter	1 in = 25.4 mm
	foot	meter	1 ft = 0.3048 m
	mile	kilometer	1 mi = 1.609 km
Area	square inch	square centimeter	1 in^2 = 6.452 cm^2
	square foot	square meter	1 ft^2 = 0.09290 m^2
Volume	cubic inch	cubic centimeter	1 in^3 = 16.39 cm^3
	cubic foot	cubic meter	1 ft^3 = 0.02832 m^3
	gallon		1 gal = 0.003789 m^3
Mass	pound-mass	kilogram	1 lbm = 0.4536 kg

Table A.1 English Units, SI Units, and Their Conversion Factors (*Continued*)

Quantity	English Units	International System* SI	Conversion Factor
	slug		1 slug = 14.59 kg
Density	slug/cubic foot	kilogram/cubic meter	1 slug/ft³ = 515.4 kg/m³
Force	pound-force	newton	1 lb = 4.448 N
Work/torque	foot-pound	newton-meter	1 ft-lb = 1.356 N·m
Pressure	pound/square inch	newton/square meter	1 psi = 6895 Pa
	pound/square foot		1 psf = 47.88 Pa
Temperature	degree Fahrenheit	degree Celsius	°F = 9/5°C + 32
	degree Rankine	kelvin	°R = 9/5K
Energy	British thermal unit	joule	1 Btu = 1055 J
	calorie		1 cal = 4.186 J
	foot-pound		1 ft-lb = 1.356 J
Power	horsepower	watt	1 hp = 745.7 W
	foot-pound/second		1 ft-lb/sec = 1.356W
Velocity	foot/second	meter/second	1 ft/sec = 0.3048 m/s
Acceleration	foot/second squared	meter/second squared	1 ft/sec² = 0.3048 m/s²
Frequency	cycle/second	hertz	1 cps = 1.000 Hz
Viscosity	pound-sec/square foot	newton-sec/square meter	1 lb-sec/ft² = 47.88 N·s/m²

*The reversed initials come from the French form of the name: Systeme International.

Table A.2 Conversions of Units

Length	Force	Mass	Velocity
1 cm = 0.3937 in	1 lb = 0.4536 kg	1 oz = 28.35 g	1 mph = 1.467 ft/sec
1 m = 3.281 ft	1 lb = 0.4448×10⁶ dyn	1 lb = 0.4536 kg	1 mph = 0.8684 knot
1 km = 0.6214 mi	1 lb = 32.17 pdl	1 slug = 32.17 lb	1 ft/sec = 0.3048 m/s
1 in = 2.54 cm	1 kg = 2.205 lb	1 slug = 14.59 kg	1 m/s = 3.281 ft/sec
1 ft = 0.3048 m	1 N = 0.2248 lb	1 kg = 2.205 lb	1 km/h = 0.278 m/s
1 mi = 1.609 km	1 dyn = 2.248×10⁻⁶ lb	1 kg = 0.06852 slug	
1 mi = 5280 ft	1 lb = 4.448 N		

Table A.2   Conversion of Units (*Continued*)

Work, Energy, and Power	Pressure	Volume
1 Btu = 778.2 ft-lb	1 psi = 2.036 in Hg	1 ft^3 = 28.32 L
1 J = 10^7 ergs	1 psi = 27.7 in H$_2$O	1 ft^3 = 7.481 gal (U.S.)
1 J = 0.7376 ft-lb	14.7 psi = 22.92 in Hg	1 gal (U.S.) = 231 in^3
1 cal = 3.088 ft-lb	14.7 psi = 33.93 ft H$_2$O	1 gal (Brit.) = 1.2 gal (U.S.)
1 cal = 0.003968 Btu	14.7 psi = 1.013 bar	1 m^3 = 1000 L
1 kWh = 3413 Btu	1 kg/cm^2 = 14.22 psi	1 ft^3 = 0.02832 m^3
1 Btu = 1.055 kJ	1 in Hg = 0.4912 psi	1 m^3 = 35.31 ft^3
1 ft-lb =1.356 J	1 ft H$_2$O = 0.4331 psi	
1 hp = 550 ft-lb/sec	1 psi = 6895 Pa	
1 hp = 0.7067 Btu/sec	1 psf = 47.88 Pa	
1 hp = 0.7455 kW	10^5 Pa = 1 bar	
1 W = 1 J/s	1 kPa = 0.145 psi	
1 W = 1.0 × 10^7 dyn-cm/s		
1 erg = 10^{-7} J		
1 quad = 10^{15} Btu		
1 therm = 10^5 Btu		
Viscosity	**Flow Rate**	
1 stoke = 10^{-4} m^2/s	1 ft^3/min = 4.719 × 10^{-4} m^3/s	
1 poise = 0.1 N · s/m^2	1 ft^3/sec = 0.02832 m^3/s	
1 lb-sec/ft^2 = 47.88 N · s/m^2	1 m^3/s = 35.31 ft^3/sec	
1 ft^2/sec = 0.0929 m^2/s	1 gal/min = 0.002228 ft^3/sec	
	1 ft^3/sec = 448.9 gal/min	

APPENDIX B

Vector Relationships

$\mathbf{A} \cdot \mathbf{B} = A_x B_x + A_y B_y + A_z B_z$

$\mathbf{A} \cdot \mathbf{B} = (A_y B_z - A_z B_y)\mathbf{i} + (A_z B_x - A_x B_z)\mathbf{j} + (A_x B_y - A_y B_x)\mathbf{k}$

gradient operator: $\nabla = \dfrac{\partial}{\partial x}\mathbf{i} + \dfrac{\partial}{\partial y}\mathbf{j} + \dfrac{\partial}{\partial z}\mathbf{k}$

divergence of $\mathbf{V} = \nabla \cdot \mathbf{V} = \dfrac{\partial u}{\partial x} + \dfrac{\partial v}{\partial y} + \dfrac{\partial w}{\partial z}$

curl of $\mathbf{V} = \nabla \times \mathbf{V} = \left(\dfrac{\partial w}{\partial y} - \dfrac{\partial v}{\partial z}\right)\mathbf{i} + \left(\dfrac{\partial u}{\partial z} - \dfrac{\partial w}{\partial x}\right)\mathbf{j} + \left(\dfrac{\partial v}{\partial x} - \dfrac{\partial u}{\partial y}\right)\mathbf{k}$

Laplace equation: $\nabla^2 \phi = 0$

Irrotational vector field: $\nabla \times \mathbf{V} = 0$

APPENDIX C

Fluid Properties

Table C.1 Properties of Water

Temperature $T\,(^\circ C)$	Density $\rho\,(\text{kg/m}^3)$	Viscosity $\mu\,(\text{N}\cdot\text{s/m}^2)$	Kinematic Viscosity $\nu\,(\text{m}^2/\text{s})$	Surface Tension $\sigma\,(\text{N/m})$	Vapor Pressure $p_v\,(\text{kPa})$	Bulk Modulus $B\,(\text{Pa})$
0	999.9	1.792×10^{-3}	1.792×10^{-6}	0.0762	0.610	204×10^7
5	1000.0	1.519	1.519	0.0754	0.872	206
10	999.7	1.308	1.308	0.0748	1.13	211
15	999.1	1.140	1.141	0.0741	1.60	214
20	998.2	1.005	1.007	0.0736	2.34	220
30	995.7	0.801	0.804	0.0718	4.24	223
40	992.2	0.656	0.661	0.0701	3.38	227
50	988.1	0.549	0.556	0.0682	12.3	230
60	983.2	0.469	0.477	0.0668	19.9	228
70	977.8	0.406	0.415	0.0650	31.2	225
80	971.8	0.357	0.367	0.0630	47.3	221
90	965.3	0.317	0.328	0.0612	70.1	216
100	958.4	0.284×10^{-3}	0.296×10^{-6}	0.0594	101.3	207×10^7

Fluid Mechanics Demystified

Table C.1E   English Properties of Water

Temperature (°F)	Density (slug/ft³)	Viscosity (lb-sec/ft²)	Kinematic Viscosity (ft²/sec)	Surface Tension (lb/ft)	Vapor Pressure (psi)	Bulk Modulus (psi)
32	1.94	3.75×10^{-5}	1.93×10^{-5}	0.518×10^{-2}	0.089	293 000
40	1.94	3.23	1.66	0.514	0.122	294 000
50	1.94	2.74	1.41	0.509	0.178	305 000
60	1.94	2.36	1.22	0.504	0.256	311 000
70	1.94	2.05	1.06	0.500	0.340	320 000
80	1.93	1.80	0.93	0.492	0.507	322 000
90	1.93	1.60	0.83	0.486	0.698	323 000
100	1.93	1.42	0.74	0.480	0.949	327 000
120	1.92	1.17	0.61	0.465	1.69	333 000
140	1.91	0.98	0.51	0.454	2.89	330 000
160	1.90	0.84	0.44	0.441	4.74	326 000
180	1.88	0.73	0.39	0.426	7.51	318 000
200	1.87	0.64	0.34	0.412	11.53	308 000
212	1.86	0.59×10^{-5}	0.32×10^{-5}	0.404×10^{-2}	14.7	300 000

Table C.2   Properties of Air at Atmospheric Pressure

Temperature T (°C)	Density ρ (kg/m³)	Viscosity μ (N·s/m²)	Kinematic Viscosity ν (m²/s)	Velocity of Sound c (m/s)
−50	1.582	1.46×10^{-5}	0.921×10^{-5}	299
−30	1.452	1.56	1.08	312
−20	1.394	1.61	1.16	319
−10	1.342	1.67	1.24	325
0	1.292	1.72	1.33	331
10	1.247	1.76	1.42	337
20	1.204	1.81	1.51	343
30	1.164	1.86	1.60	349
40	1.127	1.91	1.69	355
50	1.092	1.95	1.79	360
60	1.060	2.00	1.89	366
70	1.030	2.05	1.99	371
80	1.000	2.09	2.09	377
90	0.973	2.13	2.19	382
100	0.946	2.17	2.30	387
200	0.746	2.57	3.45	436
300	0.616	2.93×10^{-5}	4.75×10^{-5}	480

Table C.2E   English Properties of Air at Atmospheric Pressure

Temperature (°F)	Density (slug/ft^3)	Viscosity (lb-sec/ft^2)	Kinematic Viscosity (ft^2/sec)	Velocity of Sound (ft/sec)
−20	0.00280	3.34×10^{-7}	11.9×10^{-5}	1028
0	0.00268	3.38	12.6	1051
20	0.00257	3.50	13.6	1074
40	0.00247	3.62	14.6	1096
60	0.00237	3.74	15.8	1117
68	0.00233	3.81	16.0	1125
80	0.00228	3.85	16.9	1138
100	0.00220	3.96	18.0	1159
120	0.00213	4.07	18.9	1180
160	0.00199	4.23	21.3	1220
200	0.00187	4.50	24.1	1258
300	0.00162	4.98	30.7	1348
400	0.00144	5.26	36.7	1431
1000	0.000844	7.87×10^{-7}	93.2×10^{-5}	1839

Table C.3   Properties of the Standard Atmosphere

Altitude (m)	Temperature (K)	Pressure (kPa)	Density (kg/m^3)	Velocity of Sound (m/s)
0	288.2	101.3	1.225	340
500	284.9	95.43	1.167	338
1000	281.7	89.85	1.112	336
2000	275.2	79.48	1.007	333
4000	262.2	61.64	0.8194	325
6000	249.2	47.21	0.6602	316
8000	236.2	35.65	0.5258	308
10000	223.3	26.49	0.4136	300
12000	216.7	19.40	0.3119	295
14000	216.7	14.17	0.2278	295
16000	216.7	10.35	0.1665	295
18000	216.7	7.563	0.1216	295
20000	216.7	5.528	0.0889	295
30000	226.5	1.196	0.0184	302
40000	250.4	0.287	4.00×10^{-3}	317
50000	270.7	0.0798	1.03×10^{-3}	330
60000	255.8	0.0225	3.06×10^{-4}	321
70000	219.7	0.00551	8.75×10^{-5}	297
80000	180.7	0.00103	2.00×10^{-5}	269

Table C.3E   English Properties of the Atmosphere

Altitude (ft)	Temperature (°F)	Pressure (lb/ft²)	Density (slug/ft³)	Velocity of Sound (ft/sec)
0	59.0	2116	0.00237	1117
1000	55.4	2014	0.00231	1113
2000	51.9	1968	0.00224	1109
5000	41.2	1760	0.00205	1098
10000	23.4	1455	0.00176	1078
15000	5.54	1194	0.00150	1058
20000	−12.3	973	0.00127	1037
25000	−30.1	785	0.00107	1016
30000	−48.0	628	0.000890	995
35000	−65.8	498	0.000737	973
36000	−67.6	475	0.000709	971
40000	−67.6	392	0.000586	971
50000	−67.6	242	0.000362	971
100000	−51.4	23.2	3.31×10^{-5}	971

Table C.4   Properties of Ideal Gases at 300 K ($c_v = c_p - k$, $k = c_p/c_v$)

Gas	Chemical formula	Molar mass	R ft-lb / slug-°R	R kJ / kg·K	c_p ft-lb / slug-°R	c_p kJ / kg·K	k
Air		28.97	1716	0.287	6012	1.004	1.40
Argon	Ar	39.94	1244	0.2081	3139	0.5203	1.667
Carbon dioxide	CO_2	44.01	1129	0.1889	5085	0.8418	1.287
Carbon monoxide	CO	28.01	1775	0.2968	6238	1.041	1.40
Ethane	C_2H_6	30.07	1653	0.2765	10700	1.766	1.184
Helium	He	4.003	12420	2.077	31310	5.193	1.667
Hydrogen	H_2	2.016	24660	4.124	85930	14.21	1.40
Methane	CH_4	16.04	3100	0.5184	13330	2.254	1.30
Nitrogen	N_2	28.02	1774	0.2968	6213	1.042	1.40
Oxygen	O_2	32.00	1553	0.2598	5486	0.9216	1.394
Propane	C_3H_8	44.10	1127	0.1886	10200	1.679	1.12
Steam	H_2O	18.02	2759	0.4615	11150	1.872	1.33

Table C.5   Properties of Common Liquids at Atmospheric Pressure and Approximately 16°C to 21°C (60°F to 70°F)

Liquid	Specific Weight		Density		Surface Tension		Vapor Pressure	
	lb/ft^3	N/m^3	slugs/ft^3	kg/m^3	lb/ft	N/m	psia	kPa
Alcohol, ethyl	49.3	7744	1.53	789	0.0015	0.022		—
Benzene	56.2	8828	1.75	902	0.0020	0.029	1.50	10.3
Carbon tetrachloride	99.5	15 629	3.09	1593	0.0018	0.026	12.50	86.2
Glycerin	78.6	12346	2.44	1258	0.0043	0.063	2×10^{-6}	1.4×10^{-5}
Kerosene	50.5	7933	1.57	809	0.0017	0.025	—	—
Mercury*	845.5	132 800	26.29	13550	0.032	0.467	2.31×10^{-5}	1.59×10^{-4}
SAE 10 oil	57.4	9016	1.78	917	0.0025	0.036	—	—
SAE 30 oil	57.4	9016	1.78	917	0.0024	0.035	—	—
Water	62.4	9810	1.94	1000	0.0050	0.073	0.34	2.34

*In contact with air.

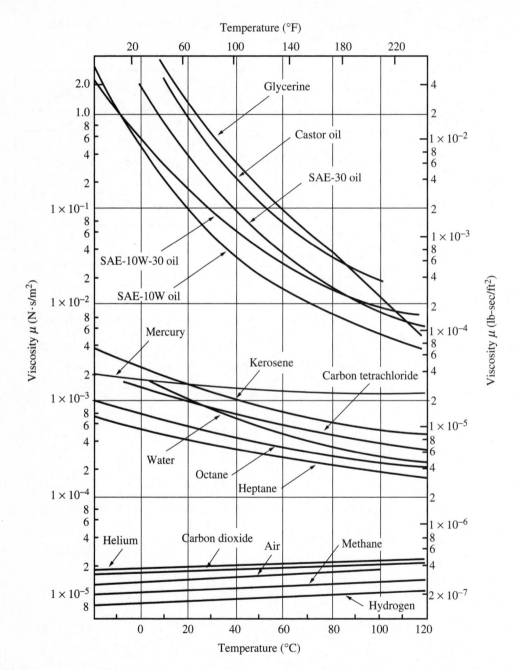

Figure C.1   Viscosity as a function of temperature. (From R.W. Fox and T.A. McDonald, *Introduction to Fluid Mechanics*, 2nd ed., John Wiley & Sons, Inc., New York, 1978.)

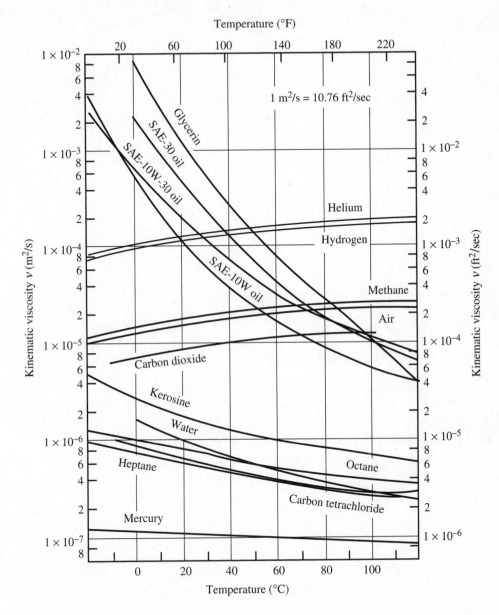

Figure C.2   Kinematic viscosity as a function of temperature, at atmospheric pressure. (From R.W. Fox and T.A. McDonald, *Introduction to Fluid Mechanics*, 2nd ed., John Wiley & Sons, Inc., New York, 1978.)

APPENDIX D

Compressible Flow Table for Air

Table D.1 Isentropic Flow

M	p/p_0	T/T_0	A/A^*	M	p/p_0	T/T_0	A/A^*
0	1.0000	1.0000	0	0.28	0.9470	0.9846	2.1656
0.04	0.9989	0.9997	14.4815	0.32	0.9315	0.9799	1.9219
0.08	0.9955	0.9987	7.2616	0.36	0.9143	0.9747	1.7358
0.12	0.9900	0.9971	4.8643	0.40	0.8956	0.9690	1.5901
0.16	0.9823	0.9949	3.6727	0.44	0.8755	0.9627	1.4740
0.20	0.9725	0.9921	2.9635	0.48	0.8541	0.9560	1.3801
0.24	0.9607	0.9886	2.4956	0.52	0.8317	0.9487	1.3034

Table D.1   Isentropic Flow (*Continued*)

M	p/p_0	T/T_0	A/A^*	M	p/p_0	T/T_0	A/A^*
0.56	0.8082	0.9410	1.2403	1.80	0.1740	0.6068	1.439
0.60	0.7840	0.9328	1.1882	1.84	0.1637	0.5963	1.484
0.64	0.7591	0.9243	1.1452	1.88	0.1539	0.5859	1.531
0.68	0.7338	0.9153	1.1097	1.90	0.1492	0.5807	1.555
0.72	0.7080	0.9061	1.0806	1.92	0.1447	0.5756	1.580
0.76	0.6821	0.8964	1.0570	1.96	0.1360	0.5655	1.633
0.80	0.6560	0.8865	1.0382	2.00	0.1278	0.5556	1.688
0.84	0.6300	0.8763	1.0237	2.04	0.1201	0.5458	1.745
0.88	0.6041	0.8659	1.0129	2.08	0.1128	0.5361	1.806
0.92	0.5785	0.8552	1.0056	2.12	0.1060	0.5266	1.869
0.96	0.5532	0.8444	1.0014	2.16	0.09956	0.5173	1.935
1.00	0.5283	0.8333	1.000	2.20	0.09352	0.5081	2.005
1.04	0.5039	0.8222	1.001	2.24	0.08785	0.4991	2.078
1.08	0.4800	0.8108	1.005	2.28	0.08251	0.4903	2.154
1.12	0.4568	0.7994	1.011	2.32	0.07751	0.4816	2.233
1.16	0.4343	0.7879	1.020	2.36	0.07281	0.4731	2.316
1.20	0.4124	0.7764	1.030	2.40	0.06840	0.4647	2.403
1.24	0.3912	0.7648	1.043	2.44	0.06426	0.4565	2.494
1.28	0.3708	0.7532	1.058	2.48	0.06038	0.4484	2.588
1.32	0.3512	0.7416	1.075	2.52	0.05674	0.4405	2.686
1.36	0.3323	0.7300	1.094	2.56	0.05332	0.4328	2.789
1.40	0.3142	0.7184	1.115	2.60	0.05012	0.4252	2.896
1.44	0.2969	0.7069	1.138	2.64	0.04711	0.4177	3.007
1.48	0.2804	0.6954	1.163	2.68	0.04429	0.4104	3.123
1.52	0.2646	0.6840	1.190	2.72	0.04165	0.4033	3.244
1.56	0.2496	0.6726	1.219	2.76	0.03917	0.3963	3.370
1.60	0.2353	0.6614	1.250	2.80	0.03685	0.3894	3.500
1.64	0.2217	0.6502	1.284	2.84	0.03467	0.3827	3.636
1.68	0.2088	0.6392	1.319	2.88	0.03263	0.3761	3.777
1.72	0.1966	0.6283	1.357	2.92	0.03071	0.3696	3.924
1.76	0.1850	0.6175	1.397	2.96	0.02891	0.3633	4.076

Table D.1   Isentropic Flow (*Continued*)

M	p/p_0	T/T_0	A/A^*	M	p/p_0	T/T_0	A/A^*
3.00	0.02722	0.3571	4.235	4.12	0.005619	0.2275	11.92
3.04	0.02564	0.3511	4.399	4.16	0.005333	0.2242	12.35
3.08	0.02416	0.3452	4.570	4.20	0.005062	0.2208	12.79
3.12	0.02276	0.3393	4.747	4.24	0.004806	0.2176	13.25
3.16	0.02146	0.3337	4.930	4.28	0.004565	0.2144	13.72
3.20	0.02023	0.3281	5.121	4.32	0.004337	0.2113	14.20
3.24	0.01908	0.3226	5.319	4.36	0.004121	0.2083	14.70
3.28	0.01799	0.3173	5.523	4.40	0.003918	0.2053	15.21
3.32	0.01698	0.3121	5.736	4.44	0.003725	0.2023	15.74
3.36	0.01602	0.3069	5.956	4.48	0.003543	0.1994	16.28
3.40	0.01512	0.3019	6.184	4.52	0.003370	0.1966	16.84
3.44	0.01428	0.2970	6.420	4.54	0.003288	0.1952	17.13
3.48	0.01349	0.2922	6.664	4.58	0.003129	0.1925	17.72
3.52	0.01274	0.2875	6.917	4.62	0.002978	0.1898	18.32
3.56	0.01204	0.2829	7.179	4.66	0.002836	0.1872	18.94
3.60	0.01138	0.2784	7.450	4.70	0.002701	0.1846	19.58
3.64	0.01076	0.2740	7.730	4.74	0.002573	0.1820	20.24
3.68	0.01018	0.2697	8.020	4.78	0.002452	0.1795	20.92
3.72	0.009633	0.2654	8.320	4.82	0.002338	0.1771	21.61
3.76	0.009116	0.2613	8.630	4.86	0.002229	0.1747	22.33
3.80	0.008629	0.2572	8.951	4.90	0.002126	0.1724	23.07
3.84	0.008171	0.2532	9.282	4.94	0.002028	0.1700	23.82
3.88	0.007739	0.2493	9.624	4.98	0.001935	0.1678	24.60
3.92	0.007332	0.2455	9.977	6.00	0.000633	0.1219	53.19
3.96	0.006948	0.2418	10.34	8.00	0.000102	0.0725	109.11
4.00	0.006586	0.2381	10.72	10.00	0.0000236	0.0476	535.94
4.04	0.006245	0.2345	11.11	∞	0	0	∞
4.08	0.005923	0.2310	11.51				

Table D.2 Normal Shock Flow

M_1	M_2	p_2/p_1	T_2/T_1	p_{02}/p_{01}
1.00	1.0000	1.0000	1.000	1.0000
1.04	0.9620	1.095	1.026	0.9999
1.08	0.9277	1.194	1.052	0.9994
1.12	0.8966	1.297	1.078	0.9982
1.16	0.8682	1.403	1.103	0.9961
1.20	0.8422	1.513	1.128	0.9928
1.24	0.8183	1.627	1.153	0.9884
1.28	0.7963	1.745	1.178	0.9827
1.30	0.7860	1.805	1.191	0.9794
1.32	0.7760	1.866	1.204	0.9758
1.36	0.7572	1.991	1.229	0.9676
1.40	0.7397	2.120	1.255	0.9582
1.44	0.7235	2.253	1.281	0.9476
1.48	0.7083	2.389	1.307	0.9360
1.52	0.6941	2.529	1.334	0.9233
1.56	0.6809	2.673	1.361	0.9097
1.60	0.6684	2.820	1.388	0.8952
1.64	0.6568	2.971	1.416	0.8799
1.68	0.6458	3.126	1.444	0.8640
1.72	0.6355	3.285	1.473	0.8474
1.76	0.6257	3.447	1.502	0.8302
1.80	0.6165	3.613	1.532	0.8127
1.84	0.6078	3.783	1.562	0.7948
1.88	0.5996	3.957	1.592	0.7765
1.92	0.5918	4.134	1.624	0.7581
1.96	0.5844	4.315	1.655	0.7395
2.00	0.5774	4.500	1.688	0.7209
2.04	0.5707	4.689	1.720	0.7022
2.08	0.5643	4.881	1.754	0.6835
2.12	0.5583	5.077	1.787	0.6649
2.16	0.5525	5.277	1.822	0.6464
2.20	0.5471	5.480	1.857	0.6281

Table D.2 Normal Shock Flow (*Continued*)

M_1	M_2	p_2/p_1	T_2/T_1	p_{02}/p_{01}
2.24	0.5418	5.687	1.892	0.6100
2.28	0.5368	5.898	1.929	0.5921
2.30	0.5344	6.005	1.947	0.5833
2.32	0.5321	6.113	1.965	0.5745
2.36	0.5275	6.331	2.002	0.5572
2.40	0.5231	6.553	2.040	0.5401
2.44	0.5189	6.779	2.079	0.5234
2.48	0.5149	7.009	2.118	0.5071
2.52	0.5111	7.242	2.157	0.4991
2.56	0.5074	7.479	2.198	0.4754
2.60	0.5039	7.720	2.238	0.4601
2.64	0.5005	7.965	2.280	0.4452
2.68	0.4972	8.213	2.322	0.4307
2.72	0.4941	8.465	2.364	0.4166
2.76	0.4911	8.721	2.407	0.4028
2.80	0.4882	8.980	2.451	0.3895
2.84	0.4854	9.243	2.496	0.3765
2.88	0.4827	9.510	2.540	0.3639
2.92	0.4801	9.781	2.586	0.3517
2.96	0.4776	10.06	2.632	0.3398
3.00	0.4752	10.33	2.679	0.3283
3.04	0.4729	10.62	2.726	0.3172
3.08	0.4706	10.90	2.774	0.3065
3.12	0.4685	11.19	2.823	0.2960
3.16	0.4664	11.48	2.872	0.2860
3.20	0.4643	11.78	2.922	0.2762

Fluid Mechanics Demystified

Table D.2 Normal Shock Flow (*Continued*)

M₁	M₂	p₂/p₁	T₂/T₁	p₀₂/p₀₁
3.24	0.4624	12.08	2.972	0.2668
3.28	0.4605	12.38	3.023	0.2577
3.30	0.4596	12.54	3.049	0.2533
3.32	0.4587	12.69	3.075	0.2489
3.36	0.4569	13.00	3.127	0.2404
3.40	0.4552	13.32	3.180	0.2322
3.44	0.4535	13.64	3.234	0.2243
3.48	0.4519	13.96	3.288	0.2167
3.52	0.4504	14.29	3.343	0.2093
3.56	0.4489	14.62	3.398	0.2022
3.60	0.4474	14.95	3.454	0.1953
3.64	0.4460	15.29	3.510	0.1887
3.68	0.4446	15.63	3.568	0.1823
3.72	0.4433	15.98	3.625	0.1761
3.76	0.4420	16.33	3.684	0.1702
3.80	0.4407	16.68	3.743	0.1645
3.84	0.4395	17.04	3.802	0.1589
3.88	0.4383	17.40	3.863	0.1536
3.92	0.4372	17.76	3.923	0.1485
3.96	0.4360	18.13	3.985	0.1435
4.00	0.4350	18.50	4.047	0.1388
4.04	0.4339	18.88	4.110	0.1342
4.08	0.4329	19.25	4.173	0.1297
4.12	0.4319	19.64	4.237	0.1254
4.16	0.4309	20.02	4.301	0.1213
4.20	0.4299	20.41	4.367	0.1173

Table D.2 Normal Shock Flow (*Continued*)

M_1	M_2	p_2/p_1	T_2/T_1	p_{02}/p_{01}
4.24	0.4290	20.81	4.432	0.1135
4.28	0.4281	21.20	4.499	0.1098
4.32	0.4272	21.61	4.566	0.1062
4.36	0.4264	22.01	4.633	0.1028
4.40	0.4255	22.42	4.702	0.9948^{-1}
4.44	0.4247	22.83	4.771	0.9628^{-1}
4.48	0.4239	23.25	4.840	0.9320^{-1}
4.52	0.4232	23.67	4.910	0.9022^{-1}
4.56	0.4224	24.09	4.981	0.8735^{-1}
4.60	0.4217	24.52	5.052	0.8459^{-1}
4.64	0.4210	24.95	5.124	0.8192^{-1}
4.68	0.4203	25.39	5.197	0.7934^{-1}
4.72	0.4196	25.82	5.270	0.7685^{-1}
4.76	0.4189	26.27	5.344	0.7445^{-1}
4.80	0.4183	26.71	5.418	0.7214^{-1}
4.84	0.4176	27.16	5.494	0.6991^{-1}
4.88	0.4170	27.62	5.569	0.6775^{-1}
4.92	0.4164	28.07	5.646	0.6567^{-1}
4.96	0.4158	28.54	5.723	0.6366^{-1}
5.00	0.4152	29.00	5.800	0.6172^{-1}
6.00	0.4042	41.83	7.941	0.2965^{-1}
7.00	0.3974	57.00	10.469	0.1535^{-1}
8.00	0.3929	74.50	13.387	0.0849^{-1}
9.00	0.3898	94.33	16.693	0.0496^{-1}
10.00	0.3875	116.50	20.388	0.0304^{-1}
∞	0.3780	∞	∞	0

Fluid Mechanics Demystified

Table D.3   Prandtl-Meyer Function

M	θ	μ	M	θ	μ
1.00	0	90.00	2.24	32.763	26.51
1.04	0.3510	74.06	2.28	33.780	26.01
1.08	0.9680	67.81	2.32	34.783	25.53
1.12	1.735	63.23	2.36	35.771	25.07
1.16	2.607	59.55	2.40	36.746	24.62
1.20	3.558	56.44	2.44	37.708	24.19
1.24	4.569	53.75	2.48	38.655	23.78
1.28	5.627	51.38	2.52	39.589	23.38
1.32	6.721	49.25	2.56	40.509	22.99
1.36	7.844	47.33	2.60	41.415	22.62
1.40	8.987	45.58	2.64	42.307	22.26
1.44	10.146	43.98	2.68	43.187	21.91
1.48	11.317	42.51	2.72	44.053	21.57
1.52	12.495	41.14	2.76	44.906	21.24
1.56	13.677	39.87	2.80	45.746	20.92
1.60	14.861	38.68	2.84	46.573	20.62
1.64	16.043	37.57	2.88	47.388	20.32
1.68	17.222	36.53	2.92	48.190	20.03
1.72	18.397	35.55	2.96	48.980	19.75
1.76	19.565	34.62	3.00	49.757	19.47
1.80	20.725	33.75	3.04	50.523	19.20
1.84	21.877	32.92	3.08	51.277	18.95
1.88	23.019	32.13	3.12	52.020	18.69
1.92	24.151	31.39	3.16	52.751	18.45
1.96	25.271	30.68	3.20	53.470	18.21
2.00	26.380	30.00	3.24	54.179	17.98
2.04	27.476	29.35	3.28	54.877	17.75
2.08	28.560	28.74	3.32	55.564	17.53
2.12	29.631	28.14	3.36	56.241	17.31
2.16	30.689	27.58	3.40	56.907	17.10
2.20	31.732	27.04	3.44	57.564	16.90

Table D.3   Prandtl-Meyer Function (*Continued*)

M	θ	μ	M	θ	μ
3.48	58.210	16.70	4.28	69.302	13.51
3.52	58.847	16.51	4.32	69.777	13.38
3.56	59.474	16.31	4.36	70.245	13.26
3.60	60.091	16.13	4.40	70.706	13.14
3.64	60.700	15.95	4.44	71.161	13.02
3.68	61.299	15.77	4.48	71.610	12.90
3.72	61.899	15.59	4.52	72.052	12.78
3.76	62.471	15.42	4.56	72.489	12.67
3.80	63.044	15.26	4.60	72.919	12.56
3.84	63.608	15.10	4.64	73.344	12.45
3.88	64.164	14.94	4.68	73.763	12.34
3.92	64.713	14.78	4.72	74.176	12.23
3.96	65.253	14.63	4.76	74.584	12.13
4.00	65.785	14.48	4.80	74.986	12.03
4.04	66.309	14.33	4.84	75.383	11.92
4.08	66.826	14.19	4.88	75.775	11.83
4.12	67.336	14.05	4.92	76.162	11.73
4.16	67.838	13.91	4.96	76.544	11.63
4.20	68.333	13.77	5.00	76.920	11.54
4.24	68.821	13.64			

Final Exams

Final Exam No. 1

1. If force, length, and time are selected as the basic dimensions (the F-L-T system), the dimensions on mass are

    (A) FL/T^2

    (B) FL^2/T

    (C) FT^2/L

    (D) FT/L^2

2. A steel needle of length L, radius r, and density ρ_s can float in water. This is possible, assuming a vertical surface-tension force, if the surface tension is greater than

    (A) $\pi r^2 L \rho_s / 2$

    (B) $\pi r^2 L \rho_s$

    (C) $2\pi r^2 L \rho_s$

    (D) $\pi r^2 L \rho_s / 4$

3. If the pressure in the water pipe is 40 kPa and $h = 20$ cm, the distance H is nearest ($S_{Hg} = 13.6$)

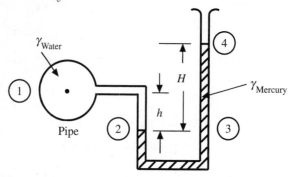

(A) 31 cm

(B) 37 cm

(C) 42 cm

(D) 49 cm

4. The gate shown is a quarter circle of radius 120 cm. The hinge is 60 cm below the water surface. The force needed to just open the 1.4-m-wide gate is nearest

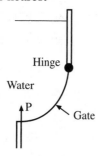

(A) 10.1 kN

(B) 12.9 kN

(C) 14.6 kN

(D) 18.2 kN

5. The velocity field is given by $\mathbf{V} = 2xy\mathbf{i} + xt\mathbf{j}$ m/s. The acceleration at the location (2, −1) at $t = 2$ is

(A) $20\mathbf{i} + \mathbf{j}$

(B) $24\mathbf{i} + 4\mathbf{j}$

(C) $8\mathbf{i} + 6\mathbf{j}$

(D) $16\mathbf{i} + 2\mathbf{j}$

6. The temperature of a fluid flow is given by $(60 - 0.2xy)°C$. The velocity field is $\mathbf{V} = 2xy\mathbf{i} + yt\mathbf{j}$ m/s. The rate of change of the temperature at $(2, -4)$ m at $t = 4$ s is

(A) $-12.8°C/s$

(B) $-10.6°C/s$

(C) $-6.4°C/s$

(D) $-4.8°C/s$

7. Select the only true statement for Bernoulli's equation:

(A) It can be applied between two points in a rotating flow.

(B) It can be applied to an unsteady flow.

(C) It can be applied to a viscous flow if the effects of viscosity are included.

(D) It can be applied between two points along a streamline in an inviscid flow.

8. Water flows at 0.01 m³/s in a pipe that splits into a 2-cm-diameter pipe and a square conduit. If the average velocity in the smaller pipe is 20 m/s, the mass flux in the conduit is nearest

(A) 3.72 kg/s

(B) 4.63 kg/s

(C) 6.41 kg/s

(D) 8.18 kg/s

9. Atmospheric air flows over the flat plate as shown. Viscosity makes the air stick to the surface creating a thin boundary layer. Estimate the average velocity $\bar{v}$ of the air across the top surface that is 10 cm above the plate if $u(y) = 800y$.

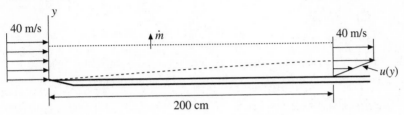

(A) 0.75 m/s

(B) 0.5 m/s

(C) 0.25 m/s

(D) 0.10 m/s

10. Water flows from a reservoir with an elevation of 25 m out a short pipe that contains a turbine to produce power. The exit is at an elevation of 5 m. If the flow rate is 0.01 m³/s, the maximum power output is nearest

  (A) 1960 W

  (B) 2430 W

  (C) 2650 W

  (D) 3480 W

11. An 88 percent efficient pump is used to increase the pressure in water in a 6-cm-diameter pipe from 120 kPa to 2000 MPa at an average velocity of 4 m/s. The required horsepower is nearest

  (A) 42.6 hp

  (B) 36.2 hp

  (C) 32.4 hp

  (D) 28.5 hp

12. A nozzle with a 2-cm-diameter outlet is attached to a 10-cm-diameter hose. If the exit velocity is 50 m/s, the force of the water on the nozzle is nearest

  (A) 9050 N

  (B) 7500 N

  (C) 2420 N

  (D) 750 N

13. A stationary blade turns a 2-cm-diameter jet through an angle of 60°. The velocity of the water is 40 m/s. The x-component of the force acting on the blade is nearest

  (A) 159 N

  (B) 251 N

  (C) 426 N

  (D) 754 N

14. Select the dimensionless group into which the variables velocity V, viscosity μ, density ρ, and radius R can be combined.

  (A) $\rho V / \mu R$

  (B) $\mu V / \rho R$

  (C) $\mu V R / \rho$

  (D) $\rho V R / \mu$

15. A proposed new design of the front portion of a ship is being studied in a water channel with a 25 to 1 scale model. A force of 4 N is measured on the model. What force would be expected on the prototype?

    (A) 12.5 kN

    (B) 24.8 kN

    (C) 48.2 kN

    (D) 62.5 kN

16. Estimate the torque needed to rotate a 10-cm-diameter, 25-cm-long cylinder inside a 10.2-cm-diameter cylinder at 800 rpm if SAE-10W oil at 20°C fills the gap. Neglect the contribution by the ends.

    (A) 1.64 N·m

    (B) 2.48 N·m

    (C) 4.22 N·m

    (D) 6.86 N·m

17. Water at 20°C is pumped through a 4-cm-diameter, 40-m-long cast-iron pipe from one reservoir to another with a water level 20 m higher. The power required of an 85 percent efficient pump for a flow rate of 0.02 m³/s is nearest

    (A) 67.9 kW

    (B) 74.8 kW

    (C) 87.2 kW

    (D) 92.5 kW

18. Estimate the depth of flow in a 2-m-wide, rectangular, finished concrete channel with a slope of 0.002 if 2 m³/s of water is flowing.

    (A) 0.45 m

    (B) 0.55 m

    (C) 0.65 m

    (D) 0.75 m

19. Estimate the percentage savings in the power needed to move a semitruck on a level surface at 70 mph through standard air with a deflector and gap seal compared to one without the deflector and seal.

    (A) 19%

    (B) 23%

    (C) 27%

    (D) 32%

20. A uniform flow $\mathbf{V} = 2\mathbf{i}$ m/s is superimposed on a doublet with strength of 2 m³/s. The radius of the resulting cylinder is

    (A) 1 m

    (B) 2 m

    (C) 3 m

    (D) 4 m

21. The maximum thickness of a laminar boundary layer that forms on a 2-m-long flat plate due to a parallel flow of water at 30°C moving at 8 m/s is

    (A) 0.98 mm

    (B) 1.12 mm

    (C) 1.86 mm

    (D) 2.44 mm

22. If the boundary layer of Prob. 21 was turbulent from the leading edge, it's thickness at the end of the plate would be nearest

    (A) 9.8 mm

    (B) 11 mm

    (C) 18 mm

    (D) 26 mm

23. A bullet is traveling at 600 m/s in standard air. Its Mach number is nearest

    (A) 1.76

    (B) 1.93

    (C) 2.01

    (D) 2.64

24. A converging nozzle is attached to a reservoir maintained at 20°C and 200 kPa. What receiver pressure would just result in $M_e = 1$?

    (A) 76 kPa

    (B) 93 kPa

    (C) 106 kPa

    (D) 264 kPa

25. Air at 25°C and 200 kPa abs enters a normal shock wave at 400 m/s. The exiting velocity is nearest

    (A) 343 m/s

    (B) 316 m/s

    (C) 303 m/s

    (D) 284 m/s

26. Air at M = 2.4 turns a convex corner of 40°. Estimate the increased Mach number after the corner.

    (A) 3.44

    (B) 3.88

    (C) 4.24

    (D) 4.98

Final Exam No. 2

1. An equation that provides a rate of flow (m³/s) in an open channel is $Q = kAR^{2/3}S^{1/2}$, where A is an area, R is a "radius." What are the units of k?

2. A viscometer consists of two 15-cm-long cylinders with radii of 5 and 4.8 cm respectively. The outer one is stationary and the inner one rotates at 150 rpm. Estimate the viscosity of the fluid if a torque of 0.05 N·m is required.

3. A vertical gate measures 120-cm high and 240-cm wide. What horizontal force at the top is needed to hold water at a depth of 1 m? A frictionless hinge exists at the bottom.

4. The U-tube is rotated about the left leg at 150 rpm. Calculate the pressure at A in the water if L is 40 cm.

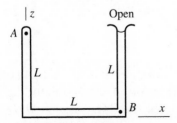

5. A velocity field is given by $\mathbf{V} = 2xy\mathbf{i} + xt\mathbf{j}$. Find a unit vector normal to the streamline at the location $(2, -1)$ at $t = 2$.

6. The density of air over a mountain varies with x and z, as do the two velocity components u and w for this steady flow. Write the expression for $D\rho/Dt$, the rate of change of the density.

7. The wind in a hurricane has a velocity of 180 km/h. Estimate the maximum force on a 1- by 2-m window.

8. Air at 20°C and 140 kPa flows in a 75- by 25-cm conduit with a flow rate of 40 m³/min. It exits from a blower at 250 kPa and 25°C in a 40-cm-diameter pipe. Calculate the average velocity in the pipe.

9. Air at 200°C and 620 kPa enters a turbine with a negligible velocity. It exits through a 4-cm-diameter pipe at 20°C and 100 kPa with a velocity of 30 m/s. Estimate the maximum power output of the turbine.

10. Water flows off a mountain into a rectangular channel at a depth of 80 cm at 20 m/s. Calculate the depth to which it is able to "jump."

11. For the 2-cm-diameter water jet shown, $V_1 = 60$ m/s, $\beta = 30°$, $V_B = 20$ m/s, and $\alpha_2 = 45°$. What blade angle α_1 should be selected for proper design?

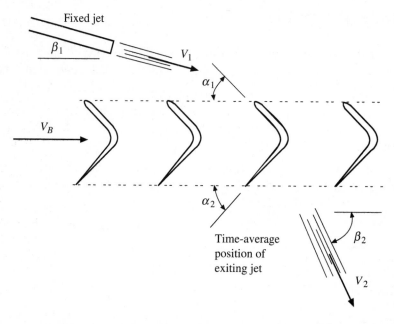

12. A disc of diameter D rotates at a distance δ from a flat plate. A liquid of density ρ and viscosity μ fills the space. Develop an expression for the torque T needed to rotate the disc at rotational speed Ω. Select ρ, Ω, D as repeating variables.

13. A study is to be performed on the flow of water around a very small organism. A larger scale model is used, 200 times larger than the organism. If water at the same temperature is used in the model study, what speed should be used to simulate a speed of 0.006 m/s by the organism?

14. Water at 15°C flows from a reservoir through a 6-cm-diameter, wrought-iron, 40-m-long pipe to the atmosphere. The pipe has a reentrant entrance and its outlet is 10 m below the water surface. Estimate the mass flux.

15. A uniform flow $V = 6\mathbf{i}$ and a doublet $\mu = 20$ m³/s are superimposed to create the flow of water around a cylinder. Calculate the minimum pressure on the cylinder if the pressure at the stagnation point is 100 kPa.

16. Water at 20°C flows parallel to a 1-m-wide by 2-m-long flat plate at 5 m/s. Estimate the drag force on one side of the plate assuming turbulent flow from the leading edge.

17. A pitot tube on a vehicle measures 5 cm of mercury in 40°C air. Estimate the speed of the vehicle.

18. A weak oblique shock with $M_1 = 2$ impacts a plane surface at 40°. Find the angle of the reflected shock.

Solutions to Quizzes and Final Exams

Chapter 1

QUIZ NO. 1

1. A $\quad \dfrac{N \cdot s}{m^2} = \dfrac{kg \cdot m}{s^2} \dfrac{s}{m^2} = \dfrac{kg}{s \cdot m}$

2. C $\quad \lambda = \dfrac{0.225 \times 4.8 \times 10^{-26}}{[100/(0.287 \times 288)] \times (3.7 \times 10^{-10})^2} = 6.5 \times 10^{-8}$ m $\quad$ or $\quad$ 65 nm

3. D $\quad h = \dfrac{p}{\rho_{Hg}g} = \dfrac{61.64 - 25}{(13.6 \times 1000) \times 9.81} = 0.275$ m $\qquad$ or $\qquad$ 275 mm

4. B $\quad p(z) = p_0 e^{-gz/RT_0} = 100 e^{-9.81 \times 6000/(287 \times 288)} = 49.1$ kPa

$\qquad \dfrac{49.1 - 47.21}{47.21} \times 100 = 4.0\%$

5. B $\quad S = \dfrac{1500/2}{1000} = 0.75$

6. D $\quad T = \tau AR \quad 0.046 = \left(\mu \dfrac{0.038 \times 120 \times 2\pi/60}{0.04 - 0.038} \right) (2\pi \times 0.038 \times 0.12) \times 0.038$

$\qquad \therefore \mu = 0.177$ N $\cdot$ s/m^2

7. A $\quad B = -\cancel{V} \left. \dfrac{\Delta p}{\Delta \cancel{V}} \right|_T \qquad 2100 = -0.1 \times \dfrac{\Delta p}{0.1 - 0.0982} \qquad \therefore \Delta p = 37.8$ MPa

8. C $\quad$ Refer to Table C.1: $p = \dfrac{4\sigma}{r} = \dfrac{4 \times 0.0736}{10 \times 10^{-6}} = 29\,400$ Pa $\qquad$ or $\qquad$ 29.4 kPa

9. D $\quad$ Since $p\cancel{V} = mRT$, and $\cancel{V}$ and m are constant, there results

$\qquad p_1/T_1 = p_2/T_2$. Hence, $370/333 = p_2/243 \qquad \therefore p_2 = 270$ kPa abs

$\qquad$ Thus, gage pressure = 170 kPa (25 psi)

10. A $\quad \cancel{V_1} = mv_1 \qquad \cancel{V_2} = mv_2 \qquad \therefore \dfrac{\cancel{V_1}}{\cancel{V_2}} = \dfrac{v_1}{v_2}$ since m is constant.

$\qquad T_2 = 295 \left(\dfrac{200}{10} \right)^{1.4-1} = 978$ K $\qquad$ or $\qquad$ 705°C

$\qquad p_2 = 100 \left(\dfrac{200}{10} \right)^{1.4} = 6630$ kPa $\qquad$ or $\qquad$ 6530 kPa gage

$\qquad$ (Assume $p_{atmosphere} = 100$ kPa.)

11. B $\quad$ Assume $T \cong 15$°C $\qquad c = \sqrt{kRT} = \sqrt{1.4 \times 287 \times 288} = 340$ m/s

$\qquad \therefore 1.5 \times 340 = 510$ m

QUIZ NO. 2

1. $F = ma$ $\qquad F = M\dfrac{L}{T^2}$ $\qquad \therefore M = \dfrac{FT^2}{L}$

2. Refer to Table C.3:

$$p = 28 + 79.48 = 107.5 \text{ kPa} \qquad h_{\text{Hg}} = \dfrac{p}{\rho g} = \dfrac{107\,500}{(13.6 \times 1000) \times 9.81} = 0.806 \text{ m}$$

3. $\tau = \mu\dfrac{du}{dr} = \mu\dfrac{10r}{16} \times 10^{-6} = 10^{-3}\dfrac{10 \times 0.004}{16} \times 10^{6} = 2.5 \text{ Pa}$

4. $T = \displaystyle\int_0^R \tau r\, dA = \int_0^{0.2} \mu\dfrac{r\Omega}{\delta} r 2\pi r\, dr = \dfrac{0.09 \times 2 \times 600 \times 2\pi/60}{0.0016} \times \dfrac{0.2^4}{4} = 2.83 \text{ N}\cdot\text{m}$

5. $h = \dfrac{4\sigma\cos\beta}{\gamma D} = \dfrac{4 \times 0.0718 \times 1.0}{(996 \times 9.81) \times 0.0002} = 0.147 \text{ m}$

6. $F = 2\sigma\pi D + \gamma_{\text{wire}}\pi D \times \pi r^2 = \pi D(2\sigma + \gamma_{\text{wire}}\pi r^2)$

7. $T = \tau AR = \mu\dfrac{\Delta u}{\delta} 2\pi RL \times R = 0.07 \times \dfrac{0.02 \times 1200 \times 2\pi/60}{0.0001} \times 2\pi \times 0.02^2 \times 2$

$\qquad = 8.8 \text{ N}\cdot\text{m} \qquad \text{Power} = \dfrac{T\Omega}{746} = \dfrac{8.8 \times 125.7}{746} = 1.5 \text{ hp}$

8. $\Delta V = \alpha_T V\Delta T = 3.8 \times 10^{-4} \times 2 \times (-10) = -0.0076 \text{ m}^3$

$$\Delta p = -\dfrac{2100\Delta V}{V} = \dfrac{2100 \times (-0.0076)}{2} = 7.98 \text{ MPa}$$

9. $pA = W \qquad 100\,000 \times 1 = 100\,000 \text{ N} \qquad \text{or} \qquad 100 \text{ kN}$

10. $T_2 = T_1\left(\dfrac{p_2}{p_1}\right)^{(k-1)/k} = (18 + 273)\left(\dfrac{100}{250 + 100}\right)^{0.4/1.4}$

$\qquad = 203 \text{ K} \qquad \text{or} \qquad -70°\text{C}$

Chapter 2

QUIZ NO. 1

1. D　$p = \gamma h$　　$9810 \times 2 = 13.6 \times 9810 h_{Hg}$

$\therefore h_{Hg} = 0.147$ m　　or　　147 mm

2. B　$p_1 = \gamma_{water} H - \gamma_{air} h = 9810 \times 0.1 = 981$ Pa

3. C　$p_w + \gamma_w h = \gamma_{Hg} H$　　$p_w = 13.6 \times 9810 \times 0.25 - 9810 \times 0.15 = 31900$ Pa

4. C　$F = \gamma h A = 9810 \times 30 \times \pi \times 0.30^2 = 83200$ N

5. A　$3P = 9810 \times \dfrac{3}{2} \times (3 \times 3) \times \dfrac{1}{3} \times 3$　　$\therefore P = 44\ 100$ N

6. C　$F = \gamma \dfrac{H}{2} \times Hw$　　$P = \gamma H \times bw$　　$F \times \dfrac{H}{3} = P \times \dfrac{b}{2}$

Moments: $\gamma \dfrac{H}{2} \times Hw \times \dfrac{H}{3} = \gamma H \times bw \times \dfrac{b}{2}$　　$\therefore H = \sqrt{3} \times 1.6 = 2.77$ m

7. D　$200 = 125 + F_B$　　$\therefore F_B = 75$ N　　$75 S = 200$　　$\therefore S = 2.67$

8. A　$\tan \alpha = \dfrac{6}{9.81} = \dfrac{H}{4.2}$　　$\therefore H = 2.57$ m　　$p = 9810 \times (H + 1.4) = 38900$ Pa

9. A　$F = p_{avg} A = \dfrac{9810}{2}(2.57 + 3.97) \times (4.2 \times 0.8) = 107\ 800$ N

10. B　$p_2 - p_1 = \dfrac{\rho \Omega^2}{2}(r_2 - r_1) = \dfrac{1000 \times (1000 \times 2\pi/60)^2}{2} \times 0.12 = 65\ 800$ Pa

11. C　$p_2 - p_1 = \dfrac{\rho \Omega^2}{2}(r_2^2 - r_1^2)$　　$p_A = \dfrac{1000 \times 10.47^2}{2} \times 0.4^2 = 8770$ Pa

where $\Omega = \dfrac{100 \times 2\pi}{60} = 10.47$ rad/s　　$p_B = \gamma L = 9810 \times 0.4 = 3920$ Pa

QUIZ NO. 2

1. $\Delta p = \gamma \Delta h$ $\qquad$ $200\,000 = 0.75 \times 9810 \times \Delta h$ $\qquad$ $\therefore \Delta h = 27.2$ m

2. $P_{in} = \rho_{in} g h = \dfrac{P_{atm}}{RT_{in}} g h = \dfrac{100}{0.287 \times 295} \times 9.81 \times 3 = 34.76$ Pa

   $P_{out} = \rho_{out} g h = \dfrac{P_{atm}}{RT_{out}} g h = \dfrac{100}{0.287 \times 253} \times 9.81 \times 3 = 40.53$ Pa $\qquad \therefore \Delta p = 5.8$ Pa

3. $p = \gamma_{water} h = \rho_{air} V^2 / 2$ $\qquad$ $9810 \times 0.25 = 0.8194 V^2 / 2$ $\qquad \therefore V = 77.4$ m/s

4. $F = \gamma h A = 9810 \times 30 \times \pi \times 0.30^2 = 83\,200$ N (The angle makes no difference.)

5. $P\left(0.8 + \dfrac{1.2}{\sin 65°}\right) = 9810 \times \dfrac{1.2}{2} \times \left(2 \times \dfrac{1.2}{\sin 65°}\right) \times \dfrac{1}{3} \times \dfrac{1.2}{\sin 65°}$

   $\therefore P = 3240$ N

6. Move F_H and F_V to the center of the circular arc (F_V produces no moment):

   $F_H = F_1 = \gamma \bar{h} A = 9810 \times 1 \times (2 \times 3) = 58\,860$ N $\qquad 2.6 P = 2 F_H$

   $\therefore P = 45\,300$ N

7. $F_B = 1200 \times 10^{-6} \times 9810 = 11.77$ N

   $\therefore W_{in\,H_2O} = 20 - F_B = 20 - 11.77 = 8.23$ N

8. The zero-pressure line slopes down from the rear toward the front:

   $\tan \alpha = \dfrac{6}{9.81} = \dfrac{H}{2}$ $\qquad \therefore H = 1.223$ m $\qquad \tan \alpha = \dfrac{6}{9.81} = \dfrac{0.223}{x}$

   $\therefore x = 0.365$ m

9. The force is the average pressure times the area:

   $F = p_{avg} A = \dfrac{\gamma}{2}(1 - 0.223) \times (2 \times 0.8) = 6100$ N

10. Use $\Omega = \dfrac{100 \times 2\pi}{60} = 10.47$ rad/s. Equate the air volume before and after:

    $\pi \times 0.6^2 \times 0.24 = \dfrac{1}{2} \pi r^2 h = \dfrac{1}{2} \pi r^2 \times \dfrac{10.47^2 r^2}{2g}$ $\qquad \therefore r = 0.419$ m

    $H_1 = \dfrac{\Omega^2 r_2^2}{2g} = \dfrac{10.47^2 \times 0.419^2}{2 \times 9.81} = 0.983$ m

$$H_2 = \frac{\Omega^2 r_2^2}{2g} = \frac{10.47^2 \times 0.6^2}{2 \times 9.81} = 2.011 \text{ m}$$

$$0.9 + 0.24 - H_1 + H_2 = 2.168 \text{ m} \qquad \therefore p_A = 9810 \times 2.168 = 21\,270 \text{ Pa}$$

Chapter 3

QUIZ NO. 1

1. C $\quad \mathbf{a} = u \dfrac{\partial \mathbf{V}}{\partial x} + v \dfrac{\partial \mathbf{V}}{\partial y} + \cancel{w \dfrac{\partial \mathbf{V}}{\partial z}} + \dfrac{\partial \mathbf{V}}{\partial t}$

$\qquad = 2yt(\mathbf{j}) + x(2t\mathbf{i}) + 2y\mathbf{i} = 12\mathbf{j} + 24\mathbf{i} + 4\mathbf{i} = 28\mathbf{i} + 12\mathbf{j} \text{ m/s}^2$

$\qquad a = \sqrt{28^2 + 12^2} = 30.5 \text{ m/s}^2$

2. D $\quad \boldsymbol{\omega} = 2\boldsymbol{\Omega} = \left(\cancel{\dfrac{\partial w}{\partial x}} - \dfrac{\partial v}{\partial y} \right)\mathbf{i} + \left(\cancel{\dfrac{\partial u}{\partial z}} - \cancel{\dfrac{\partial w}{\partial x}} \right)\mathbf{j} + \left(\dfrac{\partial v}{\partial x} - \dfrac{\partial u}{\partial y} \right)\mathbf{k}$

$\qquad = -t\mathbf{i} - 2x\mathbf{k} \qquad \therefore \boldsymbol{\omega} = -3\mathbf{i} \text{ rad/s}$

3. A $\quad a(0.5) = u\cancel{\dfrac{\partial u}{\partial x}} + \cancel{v}\dfrac{\partial u}{\partial y} + \cancel{w}\dfrac{\partial u}{\partial z} + \cancel{\dfrac{\partial u}{\partial t}} = 0 \qquad (a = 0 \text{ everywhere.})$

4. B $\quad \mathbf{V} \cdot d\mathbf{r} = 0 \text{ so } \mathbf{V} = (2xy\mathbf{i} + y^2 t\mathbf{j}) \times (dx\mathbf{i} + dy\mathbf{j}) = 0$

$\qquad xy\,dy - y^2 dx = 0 \qquad \therefore \dfrac{dx}{x} = \dfrac{dy}{y}. \text{ Integrate: } \ln x = \ln y + \ln C \qquad \therefore x = Cy$

$\qquad$ If the streamline passes through $(2, -1)$, then $C = -2$ and $x = -2y$

5. B $\quad \text{Re} = \dfrac{VD}{\nu} \cong \dfrac{0.002V}{10^{-6}} = 2000 \qquad \therefore V = 1 \text{ m/s}$

6. D $\quad$ b, g

7. C $\quad \dfrac{D\rho}{Dt} = u\dfrac{\partial \rho}{\partial x} + \cancel{v}\dfrac{\partial \rho}{\partial y} + \cancel{w}\dfrac{\partial \rho}{\partial z} + \cancel{\dfrac{\partial \rho}{\partial t}} = u\dfrac{\partial \rho}{\partial x}$

8. A $\quad$ Bernoulli's: $\dfrac{p}{\rho} = \dfrac{V^2}{2} \qquad \dfrac{0.1 \times 9810}{1.20} = \dfrac{V^2}{2} \qquad \therefore V = 40.4 \text{ m/s}$

9. C   Point 2 is inside the pitot probe: $\dfrac{V^2}{2} + \dfrac{p}{\rho} = \dfrac{\cancel{V_2^2}}{\cancel{2}} + \dfrac{p_2}{\rho}$ so $V = \sqrt{\dfrac{2}{\rho}(p_2 - p)}$

Manometer: $p + \gamma_{\text{Hg}} h = p_2 + \gamma h$ $\therefore p_2 - p = h(\gamma_{\text{Hg}} - \gamma)$

So, $V = \sqrt{\dfrac{2h}{\rho}(\gamma_{\text{Hg}} - \gamma)} = \sqrt{\dfrac{2 \times 0.1}{1000}(13.6 \times 9810 - 9810)} = 4.97 \ \text{m/s}$

10. B

11. B $-\dfrac{\Delta p}{\Delta n} = \rho \dfrac{V^2}{R}$ $-\dfrac{\Delta p}{0.02} = 1000 \times \dfrac{20^2}{0.12}$ $\therefore \Delta p = 66\ 700 \ \text{Pa}$

QUIZ NO. 2

1. $\dfrac{D\rho}{Dt} = u \dfrac{\cancel{\partial \rho}}{\cancel{\partial x}} + \cancel{v} \dfrac{\cancel{\partial \rho}}{\cancel{\partial y}} + \cancel{w} \dfrac{\partial \rho}{\partial z} + \dfrac{\cancel{\partial \rho}}{\cancel{\partial t}} = 0$

2. $a_r = v_r \dfrac{\partial v_r}{\partial r} + \dfrac{v_\theta}{r} \dfrac{\partial v_r}{\partial \theta} + v_z \dfrac{\cancel{\partial v_r}}{\cancel{\partial z}} - \dfrac{v_\theta^2}{r} + \dfrac{\cancel{\partial v_r}}{\cancel{\partial t}}$

$= \left(2 - \dfrac{8}{r^2}\right) \cancel{\cos\theta} \left(\dfrac{16}{r^3}\right) \cancel{\cos\theta} + \left(\dfrac{2}{r} + \dfrac{8}{r^3}\right) \sin\theta \left(2 - \dfrac{8}{r^2}\right) \sin\theta$

$- \dfrac{1}{r}\left(2 + \dfrac{8}{r^2}\right)^2 \sin^2\theta$

$= 0 + \left(\dfrac{2}{3} + \dfrac{8}{27}\right)\left(2 - \dfrac{8}{9}\right) - \dfrac{1}{3}\left(2 + \dfrac{8}{9}\right)^2 = -1.712 \ \text{m/s}^2$

$a_\theta = v_r \dfrac{\partial v_\theta}{\partial r} + \dfrac{v_\theta}{r} \dfrac{\partial v_\theta}{\partial \theta} + v_z \dfrac{\cancel{\partial v_\theta}}{\cancel{\partial z}} + \dfrac{v_r v_\theta}{r} + \dfrac{\cancel{\partial v_\theta}}{\cancel{\partial t}}$

$= \left(2 - \dfrac{8}{r^2}\right) \cancel{\cos\theta} \left(\dfrac{16}{r^3}\right) \sin\theta + \dfrac{1}{r}\left(2 + \dfrac{8}{r^2}\right)^2 \sin\theta \cancel{\cos\theta} - \left(\dfrac{2}{r} - \dfrac{8}{r^2}\right)$

$\cancel{\cos\theta}\left(2 + \dfrac{8}{r^2}\right)\sin\theta$

$= 0$

$a_z = 0$

3. Lagrangian: Have observers ride around in cars and record their observations.

   Eulerian: Position observers on various corners and have them record their observations.

4. At the point at $t = 2$ s, $\mathbf{V} = 4\mathbf{i} + 2\mathbf{j}$. The unit vector and the velocity vector form the product $\mathbf{V} \cdot \mathbf{n} = 0$:

$$(4\mathbf{i} + 2\mathbf{j}) \cdot (n_x \mathbf{i} + n_y \mathbf{j}) = 4n_x + 2n_y = 0 \qquad n_x^2 + n_y^2 = 1$$

$$n_x^2 + (2n_x^2)^2 = 1 \text{ and } n_x = 1/\sqrt{5}, \; n_y = 2/\sqrt{5} \qquad \therefore \mathbf{n} = (\mathbf{i} + 2\mathbf{j})/\sqrt{5}$$

5. $V \cong \dfrac{6}{40} = 0.15$ m/s $\qquad \therefore \mathrm{Re} = \dfrac{Vh}{\nu} \cong \dfrac{0.15 \times 1.2}{10^{-6}} = 1.8 \times 10^5$, a very

   turbulent flow.

6. The answer is (c). The others would (or could) be compressible (d, f), viscous (a, e), or inviscid (b, g).

7. $\mathrm{Re} = \dfrac{VL}{\nu} \qquad 3 \times 10^5 = \dfrac{2L}{1.6 \times 10^{-5}} \qquad \therefore L = 2.4$ m

8. $\dfrac{V_2^2}{2} + \dfrac{p_2}{\rho} + \cancel{gh_2} = \dfrac{V_1^2}{2} + \dfrac{p_1}{\rho} + \cancel{gh_1} \qquad$ Units: $\dfrac{\mathrm{N/m^2}}{\mathrm{kg/m^3}} = \dfrac{(\mathrm{kg \cdot m/s^2})/m^2}{\mathrm{kg/m^3}} = \dfrac{\mathrm{m^2}}{\mathrm{s^2}}$

   $\dfrac{240\,000}{1000} = \dfrac{V_1^2}{2} + \dfrac{112\,000}{1000} \qquad \therefore V_1 = 16$ m/s

9. $\dfrac{V_1^2}{2} + \dfrac{p_1}{\rho} = \dfrac{V_2^2}{2} + \dfrac{\cancel{p_2}}{\cancel{\rho}} \qquad \therefore p_1 = \rho \dfrac{(4V)^2}{2} - \rho \dfrac{V^2}{2} = \rho \dfrac{15V^2}{2}$

   Just inside the pitot tube: $p_T = \rho(4V)^2/2 = 8\rho V^2$

   $$p_1 + \gamma_{Hg} h = p_T + \gamma h \quad \text{so that} \quad 8\rho V^2 - \rho \dfrac{15V^2}{2} = h(\gamma_{Hg} - \gamma)$$

   $$V = \sqrt{\dfrac{2 \times 0.2}{1000}(13.6 \times 9810 - 9810)} = 7.03 \text{ m/s}$$

10. $p = \dfrac{1}{2}\rho V^2 \cong \dfrac{1}{2} \times 1.2 \times (120\,000/3600)^2 = 667$ Pa

    $F = pA = 667 \times \pi \times 0.2^2 = 83.8$ N

Chapter 4

QUIZ NO. 1

1. B

2. C $\rho = \dfrac{340}{0.287 \times 298} = 3.975$ kPa

$\therefore \dot{m} = 3.975 \times \pi \times 0.05^2 \times 40 = 1.25$ kg/s

3. D $A_1 V_1 = A_2 V_2 \qquad 0.02 \times 0.04 \times 16 = \pi \times 0.03^2 V_2 \qquad \therefore V_2 = 4.53$ m/s

4. B $\dfrac{d\cancel{V}}{dt} = \dfrac{d}{dt}\left(\dfrac{4}{3}\pi R^3\right) = 4\pi R^2 \dfrac{dR}{dt} = \dfrac{\pi}{2} D^2 \dfrac{dD}{dt}$

$0.01 = \dfrac{\pi}{2} \times 0.5^2 \times \dfrac{dD}{dt} \qquad \therefore \dfrac{dD}{dt} = 0.0255$ m/s

5. A $1000 \times \pi \times 0.02^2 \times V_1 = 1000 \times 0.002 + 2.5 \qquad \therefore V_1 = 3.58$ m/s

6. B It applies to compressible flow as well as incompressible flow.

7. D $V_2 = 10 \times \dfrac{8^2}{4^2} = 40$ m/s $\qquad p_1 = \cancel{p_2} + 1000 \times \left(\dfrac{40^2 - 10^2}{2}\right) = 750\,000$ Pa

8. C The two surfaces are sections 1 and 2. $\qquad 135 = 25 + 20 \times \dfrac{V^2}{2 \times 9.81}$

$\therefore V = 10.39$ m/s $\qquad$ and $\qquad Q = 10.39 \times \pi \times 0.12^2 = 0.470$ m³/s

9. A $Q = A_1 V_1 = \pi \times 0.05^2 \times 10 = 0.0854$ m³/s $\qquad V_2 = 10 \times \dfrac{10^2}{20^2} = 2.5$ m/s

$-\dfrac{\dot{W}_S}{\dot{m}g} = \dfrac{V_2^2}{2g} + \dfrac{\cancel{p_2}}{\cancel{\gamma_2}} + \cancel{z_2} - \dfrac{p_1}{\gamma_1} - \dfrac{V_1^2}{2g} - \cancel{z_1} + \cancel{h_L}$

$-\dfrac{\dot{W}_S}{1000 \times 0.0854} = \dfrac{2.5^2 - 10^2}{2} - \dfrac{800\,000}{1000} \qquad \therefore \dot{W}_S = 72\,300$ W

$\dot{W}_T = \eta_T \dot{W}_S = 0.9 \times 72.3 = 65.1$ kW

10. D $0.04 = \pi \times 0.05^2 \times V_1 = \pi \times 0.02^2 \times V_2$

$\therefore V_1 = 5.09$ m/s   and $V_2 = 31.8$ m/s

$\dfrac{V_1^2}{2} + \dfrac{p_1}{\rho} = \dfrac{V_2^2}{2}$ $\dfrac{5.09^2}{2} + \dfrac{p_1}{1000} = \dfrac{31.8^2}{2}$ $\therefore p_1 = 490\,000$ Pa

$p_1 A_1 - F_N = \dot{m}(V_2 - V_1)$

$490\,000 \times \pi \times 0.05^2 - F_N = 1000\pi \times 0.05^2 \times 5.09 \times (31.8 - 5.09)$

$\therefore F_N = 2780$ N

11. B Continuity: $V_2 = V_1 \dfrac{y_1}{y_2} = 10 \times \dfrac{0.5}{y_2} = \dfrac{5}{y_2}$

Momentum: $F_1 - F_2 = \dot{m}(V_2 - V_1)$ Use $F = \gamma h_c A$

$9810 \times 0.25 \times 0.5 w - 9810 \times \dfrac{y_2}{2} \times y_2 w = 1000 \times 0.5 w \times 10 \times \left(\dfrac{5}{y_2} - 10 \right)$

$0.25 - y_2^2 = 10.19 \times \left(\dfrac{0.5 - y_2}{y_2} \right)$ or $0.5 y_2 + y_2^2 = 10.19$

$\therefore y_2 = 2.95$ m

12. C $-F = \dot{m}_r (\cancel{V_{r2x}} - V_{r1x})$ $F = \dot{m}_r V_{r1} = 1000 \times \pi \times 0.03^2 \times 20^2 = 1131$ N

13. A $\left. \begin{array}{l} 40 \sin 30° = V_{r1} \sin \alpha_1 \\ 40 \cos 30° = 20 + V_{r1} \cos \alpha_1 \end{array} \right\}$ $\therefore V_{r1} = 24.79$ m/s $\alpha_1 = 53.8°$

14. B $\left. \begin{array}{l} V_2 \sin \beta_2 = 24.79 \sin 45° \\ V_2 \cos \beta_2 = 20 - 24.79 \cos 45° \end{array} \right\}$ $\therefore V_2 = 17.7$ m/s $\beta_2 = 81.98°$

$-F = 1000 \times \pi \times 0.01^2 \times 40 \times (17.7 \cos 81.98° - 40 \cos 30°) = -404$ N

QUIZ NO. 2

1. $A_1 V_1 = A_2 V_2$ $0.02 \times 0.04 \times 16 = \pi \times 0.03^2 V_2$ $\therefore V_2 = 4.53$ m/s

2. $5 = \dfrac{240}{0.287 \times 413} \times 0.2 \times 0.2 \times V_2$ $\therefore V_2 = 61.7$ m/s

3. $\rho_1 A_1 V_1 = \rho_2 A_2 V_2 \qquad \therefore \frac{p_1}{RT_1}\pi\frac{d_1^2}{4}V_1 = \rho_2\pi\frac{d_2^2}{4}V_2 \qquad \therefore V_2 = \frac{d_1^2 p_1}{\rho_2 d_2^2 RT_1}V_1$

$$V_2 = \frac{0.32^2 \times 350}{3.5 \times 0.20^2 \times 0.287 \times 313} \times 10 = 28.5 \text{ m/s}$$

4. Assume $\rho \cong 1.22 \text{ kg/m}^3$. For $u = 40$ m/s, $800y = 40$. $\therefore y = 0.05$ m. Continuity gives

$$1.22 \times (0.1 \times 1.2) \times 40 = \dot{m} + 1.22 \times (0.05 \times 1.2) \times 20 + 1.22 \times (0.05 \times 1.2) \times 40$$

$$\therefore \dot{m} = 1.46 \text{ kg/s}$$

5. $\dfrac{dm}{dt} = 1000 \times \pi \times 0.02^2 \times 5 - 2 - 2.5 \qquad \therefore \dfrac{dm}{dt} = 2.33 \text{ kg/s}$

6. $-\dfrac{\cancel{\dot{W}_S}}{\cancel{m}g} = \dfrac{V_2^2}{2g} + z_2 + \dfrac{\cancel{p_2}}{\cancel{\gamma_2}} - \dfrac{\cancel{V_2^2}}{\cancel{2g}} - z_1 - \dfrac{\cancel{p_1}}{\cancel{\gamma_1}} + K\dfrac{V^2}{2g} \qquad V = \dfrac{A_2}{A}V_2 = \dfrac{16}{144}V_2$

$$0 = \frac{(9V)^2}{2g} + 10 + \frac{\cancel{p_2}}{\cancel{\gamma_2}} - \frac{\cancel{V_2^2}}{\cancel{2g}} - 25 - \frac{\cancel{p_1}}{\cancel{\gamma_1}} + 2\frac{V^2}{2g}$$

$$\therefore V = 1.88 \text{ m/s} \qquad \text{and} \qquad Q = \pi \times 0.06^2 \times 1.88 = 0.0213 \text{ m}^3\text{/s}$$

7. $Q = A_1 V_1 = 0.25 \times 3.5 \times 2.2 = 1.925 \text{ m}^3\text{/s}$. For maximum power output, $V_2 \cong 0$.

$$-\frac{\dot{W}_S}{\dot{m}g} = \frac{\cancel{V_2^2}}{\cancel{2g}} + \frac{\cancel{p_2}}{\cancel{\gamma_2}} + \cancel{z_2} - \frac{\cancel{p_1}}{\cancel{\gamma_1}} - \frac{\cancel{V_1^2}}{\cancel{2g}} - z_1 + \cancel{h_L}$$

$$\dot{W}_S = \dot{m}gz_1 = (1000 \times 1.925) \times 9.81 \times 10 = 189\,000 \text{ W}$$

$$\dot{W}_T = \eta_T \dot{W}_S = 0.88 \times 189 = 166 \text{ kW}$$

8. $-\dfrac{\dot{W}_S}{\dot{m}} = \dfrac{p_2}{\rho} - \dfrac{p_1}{\rho} \qquad \therefore \dot{W}_S = \dot{m}\dfrac{p_1 - p_2}{\rho} = 0.020\rho \times \dfrac{120 - 800}{\rho} = -13.6 \text{ kW}$

$$\dot{W}_P = \frac{13.6}{0.85} = 16 \text{ kW} \qquad \text{or} \qquad 16/0.746 = 21.4 \text{ hp}$$

9. $\dot{m} = \rho_2 A_2 V_2 = \dfrac{500}{0.287 \times 433} \times \pi \times 0.01^2 \times 200 = 0.253 \text{ kg/s}$

$$\frac{\dot{Q} - \dot{W}_S}{\dot{m}} = \frac{V_2^2 - \cancel{V_1^2}}{2} + \frac{p_2}{\rho_2} - \frac{p_1}{\rho_1}$$

$$\therefore \frac{\dot{Q} - (-18\,000)}{0.253} = \frac{200^2}{2} + 287 \times 433 - 287 \times 298 \qquad \therefore \dot{Q} = -3140 \text{ J/s}$$

10. Static pressure = total pressure: $p_1 = p_2 + \rho \dfrac{V_2^2}{2}$

Energy: $\dfrac{-\dot{W}_S}{\dot{m}} = \dfrac{V_2^2 - V_1^2}{2} + \dfrac{p_2}{\rho_2} - \dfrac{p_1}{\rho_1} = -\dfrac{V_1^2}{2}$

$\therefore \dot{W}_T = \dot{m} \dfrac{V_1^2}{2} \eta_T = 1000 \times \pi \times 0.12^2 \times 20 \times \dfrac{20^2}{2} \times 0.9 = 163\,000$ W

11. Continuity: $V_2 = 9V_1$ Energy: $\dfrac{V_1^2}{2} + \dfrac{200\,000}{1000} = \dfrac{(9V_1)^2}{2}$

$\therefore V_1 = 2.236$ m/s   and $V_2 = 20.12$ m/s

$\dot{m} = \rho A V = 1000 \times \pi \times 0.015^2 \times 2.236 = 1.581$ kg/s

Momentum: $p_1 A_1 - R_x = \dot{m}(\cancel{V_{2x}} - V_{1x})$ $R_y = \dot{m}(V_{2y} - \cancel{V_{1y}})$

x-direction: $200\,000 \times \pi \times 0.015^2 - R_x = 1.581 \times (-2.236)$ $\therefore R_x = 145$ N

y-direction: $R_y = 1.581 \times 20.12 = 31.8$ N

$R_{\text{total}} = \sqrt{145^2 + 31.8^2} = 148$ N

12. Continuity: $y_2 = y_1 \dfrac{V_1}{V_2} = 0.6 \times 4 = 2.4$ m

Momentum: $F_1 - F_2 = \dot{m}(V_2 - V_1)$. Use $F = \gamma h_c A$.

$9810 \times 0.3 \times 0.6w - 9810 \times 1.2 \times 2.4w = 1000 \times 2.4w \times V_2 \times (V_2 - 4V_2)$

$\therefore V_2 = 1.918$ m/s   and $V_1 = 7.672$ m/s

13. $F = \rho A_1 V_{r1}(V_{r1x} - V_{r2x}) = 1000 \times \pi \times 0.02^2 \times 70(70 - 70\cos 30°) = 825$ N

14. $\left.\begin{array}{l} 40\sin 20° = V_{r1}\sin\alpha_1 \\ 40\cos 20° = 15 + V_{r1}\cos\alpha_1 \end{array}\right\}$ $\therefore V_{r1} = 26.41$ m/s $\alpha_1 = 31.2°$

$\left.\begin{array}{l} V_2 \sin\beta_2 = 26.41\sin 50° \\ V_2 \cos\beta_2 = 15 - 26.41\cos 50° \end{array}\right\}$ $\therefore V_2 = 20.3$ m/s $\beta_2 = 95.6°$

$-F = 1000 \times \pi \times 0.01^2 \times 40 \times (-20.3\sin 5.6° - 40\cos 20°) = -497$ N

$\dot{W} = F \times V_B = 497 \times 15 = 7455$ W   or   10 hp

Chapter 5

QUIZ NO. 1

1. A $\dfrac{\partial u}{\partial x} = -\dfrac{\partial v}{\partial y}$ $\quad \therefore \dfrac{\partial v}{\partial y} = -\dfrac{\partial u}{\partial x} = -\dfrac{\partial (Ay)}{\partial x} = 0$ $\quad$ Thus, $v(x,y) = f(x)$

But, $v(x,0) = 0$ $\quad$ so $\quad$ $f(x) = 0$ $\quad$ and $\quad$ $v(x,y) = 0$.

2. B $\dfrac{\partial u}{\partial x} + \dfrac{\partial v}{\partial y} = 0$. Hence, $\dfrac{\partial v}{\partial y} = 0$ implies that it is possible for $v = f(x)$.

3. C $\dfrac{D\rho}{Dt} = -\rho\left(\dfrac{\partial u}{\partial x} + \dfrac{\partial v}{\partial y}\right) = -2(16x + 8x) = -2(16 \times 1 + 8 \times 1) = -48 \text{ kg/m}^3\text{/s}$

4. C $\dfrac{\partial v}{\partial y} = -\dfrac{\partial u}{\partial x} = -\dfrac{2(x^2 + y^2) - 2x(2x)}{(x^2 + y^2)^2} = \dfrac{2(x^2 - y^2)}{(x^2 + y^2)^2}$

$\therefore v(x,y) = \int \dfrac{2(x^2 - y^2)}{(x^2 + y^2)^2} dy = -\dfrac{2y}{x^2 + y^2} + f(x)$ $\quad f(x) = 0$ since $v(x, 0) = 0$

5. D $\dfrac{1}{r}\dfrac{\partial}{\partial r}(rv_r) = -\dfrac{1}{r}\dfrac{\partial v_\theta}{\partial \theta} = -\dfrac{1}{r}\left(25 + \dfrac{1}{r^2}\right)\sin\theta$

$\therefore rv_r = -\int\left(25 + \dfrac{1}{r^2}\right)\sin\theta\, dr = -\left(25r - \dfrac{1}{r}\right)\sin\theta + f(\theta)$

$v_r(r,0) = 0$ $\quad$ so $\quad$ $f(\theta) = 0$ $\quad$ $\therefore v_r(r,\theta) = -(25 - 1/r^2)\sin\theta$

6. A $\rho\left(u\dfrac{\partial u}{\partial x} + \cancel{v}\dfrac{\partial u}{\partial y}\right) = -\dfrac{\partial p}{\partial x} + \mu\left(\cancel{\dfrac{\partial^2 u}{\partial x^2}} + \dfrac{\partial^2 u}{\partial y^2} + \cancel{\dfrac{\partial^2 u}{\partial z^2}}\right)$ $\quad \therefore \dfrac{\partial p}{\partial x} = \mu\dfrac{\partial^2 u}{\partial y^2}$

QUIZ NO. 2

1. The differential continuity equation $\dfrac{D\rho}{Dt} + \rho\nabla \cdot \mathbf{V} = 0$ simplifies to

$\cancel{\dfrac{\partial \rho}{\partial t}} + u\dfrac{\partial \rho}{\partial x} + v\cancel{\dfrac{\partial \rho}{\partial y}} + w\cancel{\dfrac{\partial \rho}{\partial z}} + \rho\left(\dfrac{\partial u}{\partial x} + \cancel{\dfrac{\partial v}{\partial y}} + \cancel{\dfrac{\partial w}{\partial z}}\right) = u\dfrac{\partial \rho}{\partial x} + \rho\dfrac{\partial u}{\partial x} = 0$

2. (a) $\dfrac{\partial u}{\partial x} \cong \dfrac{\Delta u}{\Delta x} = \dfrac{52-60}{0.4} = -20$ $\therefore \dfrac{\partial p}{\partial x} = -\dfrac{\rho}{u}\dfrac{\partial u}{\partial x} = \dfrac{1.2}{60} \times 20 = 0.4 \text{ kg/m}^4$

    (b) $\dfrac{\partial u}{\partial x} \cong \dfrac{\Delta u}{\Delta x} = \dfrac{60-64}{0.4} = -10$ $\therefore \dfrac{\partial p}{\partial x} = -\dfrac{\rho}{u}\dfrac{\partial u}{\partial x} = \dfrac{1.2}{60} \times 10 = 0.2 \text{ kg/m}^4$

3. $\dfrac{\partial v}{\partial y} = -\dfrac{\partial u}{\partial x} \cong \dfrac{8.2-11.1}{0.016} = -181.2$ $\therefore v_B \cong -\dfrac{\partial u}{\partial x}\Delta y = -181.2 \times 0.002$

    $= -0.36 \text{ m/s}$

4. $\dfrac{1}{r}\dfrac{\partial}{\partial r}(rv_r) + \dfrac{1}{r}\cancel{\dfrac{\partial v_\theta}{\partial \theta}} + \cancel{\dfrac{\partial v_z}{\partial z}} = 0$ $\therefore \dfrac{\partial}{\partial r}(rv_r) = 0$ so $rv_r = C$ $\therefore v_r = \dfrac{C}{r}$

5. $\dfrac{1}{r}\dfrac{\partial}{\partial r}(rv_r) = -\dfrac{1}{r}\dfrac{\partial v_\theta}{\partial \theta} = \dfrac{1}{r}\left(25 + \dfrac{1}{r^2}\right)\cos\theta$

    $\therefore rv_r = \int\left(25 + \dfrac{1}{r^2}\right)\cos\theta\, dr = \left(25r - \dfrac{1}{r}\right)\cos\theta + f(\theta)$

    But, $v_r(r, 90°) = 0$ so $f(\theta) = 0$ $\therefore v_r(r,\theta) = (25 - 1/r^2)\cos\theta$

6. $\rho\cancel{v_r}\dfrac{\partial v_z}{\partial r} + \cancel{\dfrac{v_\theta}{r}\dfrac{\partial v_z}{\partial \theta}} + v_z\cancel{\dfrac{\partial v_z}{\partial z}} + \cancel{\dfrac{\partial v_z}{\partial t}} = -\dfrac{\partial p}{\partial z} + \rho g_z$

    $+\mu\dfrac{\partial^2 v_z}{\partial r^2} + \dfrac{1}{r}\dfrac{\partial v_z}{\partial r} + \dfrac{1}{r^2}\cancel{\dfrac{\partial^2 v_z}{\partial \theta^2}} + \cancel{\dfrac{\partial^2 v_z}{\partial z^2}}$

    $\therefore \dfrac{\partial p}{\partial z} = \rho g_z + \mu\dfrac{\partial^2 v_z}{\partial r^2} + \dfrac{1}{r}\dfrac{\partial v_z}{\partial r}$

Chapter 6

QUIZ NO. 1

1. D $[\rho] = \dfrac{M}{L^3} = \dfrac{FT^2/L}{L^3} = \dfrac{FT^2}{L^4}$

2. C $\dfrac{1}{T}, \quad \dfrac{M}{LT}, \quad L, \quad \dfrac{M}{L^3} \qquad \dfrac{1}{T} \times \dfrac{LT}{M} \times \dfrac{M}{L^3} \times L^2 \qquad \therefore \pi = \dfrac{\omega\rho d^2}{\mu}$

3. C   Viscosity is the only variable with M as one of its dimensions.

4. A $\dfrac{L}{T}$, $\dfrac{M}{L^3}$, L, $\dfrac{L}{T^2}$. The density cannot enter since there is only one M.

The variables combine as V^2/gH, so $V = C\sqrt{gH}$.

5. B   Select repeating variables V, ρ, H. Dimensions $\dfrac{L}{T}$, $\dfrac{M}{L^3}$, L, $\dfrac{L}{T^2}$, $\dfrac{M}{LT}$

$$\pi_1 = \dfrac{V}{\sqrt{g}\sqrt{H}} \qquad \pi_2 = \dfrac{\rho\sqrt{g}H^{3/2}}{\mu} \qquad \therefore \dfrac{V}{\sqrt{gH}} = f\left(\dfrac{\rho\sqrt{g}H^3}{\mu}\right)$$

6. D $F_L = f(V, L, c, \alpha, \rho)$. Repeating variables V, ρ, L.

$$\pi_1 = \dfrac{F_L}{\rho V^2 c^2} \qquad \pi_2 = \dfrac{c}{L} \qquad \pi_3 = \alpha \qquad \therefore \dfrac{F_L}{\rho V^2 c^2} = f\left(\dfrac{c}{L}, \alpha\right)$$

7. B $\dfrac{V_m l_m}{\nu_m} = \dfrac{V_p l_p}{\nu_p} \qquad \therefore V_m = V_p \dfrac{l_p}{l_m} = 90 \times 5 = 450$ km/hr

8. A $\mathrm{Fr}_p = \mathrm{Fr}_m \qquad \dfrac{V_p^2}{l_p g_p} = \dfrac{V_m^2}{l_m g_m}$

$$\therefore V_m = V_p \times \sqrt{\dfrac{l_m}{l_p} \times \dfrac{g_m}{g_p}} = 12 \times \sqrt{\dfrac{1}{10}} \times 1 = 3.79 \text{ m/s}$$

9. C $\mathrm{Fr}_p = \mathrm{Fr}_m \qquad \dfrac{V_p^2}{l_p g_p} = \dfrac{V_m^2}{l_m g_m} \qquad \therefore \left(\dfrac{V_p}{V_m}\right)^2 = \dfrac{l_p}{l_m}$

$$Q_m = Q_p \times \left(\dfrac{l_p}{l_m}\right)^2 \times \dfrac{V_p}{V_m} = 1.8 \times 10^2 \times \sqrt{10} = 569 \text{ m}^3\text{/s}$$

10. B $F_p = F_m \times \dfrac{\rho_p}{\rho_m} \times \left(\dfrac{l_p}{l_m}\right)^2 \times \left(\dfrac{V_p}{V_m}\right)^2 = 20 \times 1 \times 10^2 \times 10 = 20\,000$ N

QUIZ NO. 2

1. $\mu = \tau \dfrac{dy}{du}$ $[\mu] = [\tau]\left[\dfrac{dy}{du}\right] = \dfrac{F}{L^2} \times \dfrac{L}{L/T} = \dfrac{FT}{L^2}$

2. $\dfrac{ML^2}{T^3}$, L, $\dfrac{L}{T}$, $\dfrac{M}{LT^2}$ $\dfrac{ML^2}{T^3} \times \dfrac{LT^2}{M} \times \dfrac{T}{L} \times \dfrac{1}{L^2}$ $\therefore \pi = \dfrac{\dot{W}}{\Delta p \, V d^2}$

3. $[V] = \dfrac{L}{T}$ $[g] = \dfrac{L}{T^2}$ $[h] = L$ $[\rho] = \dfrac{M}{L^3}$

Since M occurs in only one variable, that variable ρ cannot be included in the relationship. The remaining three terms are combined to form a single π-term; it is formed by observing that T occurs in only two of the variables, thus V^2 is in the numerator, and g is in the denominator. The length dimension is then canceled by placing h in the denominator. The single π term is $\pi = V^2/gh$. Since this π term depends on all other π terms and there are none, it must be at most a constant. Hence, we conclude that

$V = C\sqrt{gh}$

4. $[V, w, l, g, \rho, \mu] = \dfrac{L}{T}, L, L, \dfrac{L}{T^2}, \dfrac{M}{L^3}, \dfrac{M}{LT}$

$\pi_1 = \dfrac{V\rho w}{\mu}$ $\pi_2 = \dfrac{w}{l}$ $\pi_3 = \dfrac{V^2}{wg}$ $\therefore \dfrac{V\rho w}{\mu} = f\left(\dfrac{w}{l}, \dfrac{V^2}{wg}\right)$

5. $[\Delta p, d, V, \mu, \rho, \varepsilon, L] = \dfrac{M}{LT^2}, L, \dfrac{L}{T}, \dfrac{M}{LT}, \dfrac{M}{L^3}, L, L$

$\pi_1 = \dfrac{\Delta p}{\rho V^2}$ $\pi_2 = \dfrac{V d \rho}{\mu}$ $\pi_3 = \dfrac{d}{\varepsilon}$ $\pi_4 = \dfrac{d}{L}$ $\therefore \dfrac{\Delta p}{\rho V^2} = f\left(\dfrac{V d \rho}{\mu}, \dfrac{d}{\varepsilon}, \dfrac{d}{L}\right)$

6. $[F_D, d, V, \mu, \rho, g] = \dfrac{ML}{T^2}, L, \dfrac{L}{T}, \dfrac{M}{LT}, \dfrac{M}{L^3}, \dfrac{L}{T^2}$

$\pi_1 = \dfrac{F_D}{\rho d^2 V^2}$ $\pi_2 = \dfrac{V d \rho}{\mu}$ $\pi_3 = \dfrac{V^2}{gd}$ $\therefore \dfrac{F_D}{\rho d^2 V^2} = f\left(\dfrac{V d \rho}{\mu}, \dfrac{V^2}{gd}\right)$

7. $\mathrm{Re}_p = \mathrm{Re}_m$ $\dfrac{V_p l_p}{\nu_p} = \dfrac{V_m l_m}{\nu_m}$ $\therefore V_m = V_p \times \dfrac{l_p}{l_m} \times \dfrac{\nu_m}{\nu_p} = 50 \times \dfrac{1}{10} \times 1 = 5 \text{ m/s}$

8. $Fr_p = Fr_m$ $\qquad \dfrac{V_p^2}{l_p g_p} = \dfrac{V_m^2}{l_m g_m}$

$$\therefore V_m = V_p \times \sqrt{\dfrac{l_m}{l_p}} \times \dfrac{g_m}{g_p} = 10 \times \sqrt{\dfrac{1}{40}} \times 1 = 1.58 \text{ m/s}$$

$$F_p = F_m \times \dfrac{\rho_p}{\rho_m} \times \left(\dfrac{l_p}{l_m}\right)^2 \times \left(\dfrac{V_p}{V_m}\right)^2 = 15 \times 1 \times 40^2 \times 1.58^2 = 60\,000 \text{ N}$$

9. $M_p = M_m$ $\qquad V_p = V_m \dfrac{c_p}{c_m} = 250 \times \dfrac{325}{340} = 239 \text{ m/s}$ (See Table C.3)

$$F_p = F_m \times \dfrac{\rho_p}{\rho_m} \times \left(\dfrac{l_p}{l_m}\right)^2 \times \left(\dfrac{V_p}{V_m}\right)^2 = 200 \times \dfrac{0.8194}{1.225} \times 20^2 \times \left(\dfrac{239}{250}\right)^2 = 48\,900 \text{ N}$$

Chapter 7

QUIZ NO. 1

1. A $\quad 2000 = \dfrac{V \times 0.02}{1.8 \times 10^{-5}}$ $\qquad \therefore V = 1.8 \text{ m/s}$

2. D $\quad V = \dfrac{Q}{A} = \dfrac{0.0002}{\pi \times 0.01^2} = 0.637 \text{ m/s}$

$\qquad L_E = 0.065 \times D \times \dfrac{VD}{\nu} = 0.065 \times 0.02^2 \times \dfrac{0.637}{0.661 \times 10^{-6}} = 25.1 \text{ m}$

3. B $\quad V = \dfrac{Q}{A} = \dfrac{0.006 / 60}{\pi \times 0.02^2} = 0.0796 \text{ m/s}$

$\qquad \Delta p = \dfrac{8 \mu L V}{r_0^2} = \dfrac{8 \times 0.656 \times 10^{-3} \times 20 \times 0.0796}{0.02^2} = 20.9 \text{ Pa}$

$\qquad \text{Re} = \dfrac{0.0796 \times 0.02}{0.661 \times 10^{-6}} = 2410.$ $\quad$ May not be laminar.

4. C $\quad 2000 = \dfrac{V \times 0.012}{1.007 \times 10^{-6}}$ $\qquad \therefore V = 0.1678 \text{ m/s}$

$$\cancel{\Delta p} + \gamma \Delta h = \frac{8\mu LV}{r_0^2} \qquad 9810 \times \frac{\Delta h}{L} = \frac{8 \times (1.005 \times 10^{-3}) \times 0.1678}{0.006^2}$$

$$\sin \alpha = \frac{\Delta h}{L} = 0.00382 \qquad \therefore \alpha = 0.219° \text{ downward}$$

5. B $\quad u(y) = \frac{1}{2\mu} \frac{dp}{dx}(y^2 - by) + \frac{U}{b}y \qquad \tau = \mu \frac{du}{dy} = \frac{1}{2} \frac{dp}{dx}(2y - b) + \frac{U}{b}$

$$0 = \frac{1}{2} \frac{dp}{dx}(-b) + \frac{U}{b} \qquad \therefore \frac{dp}{dx} = \frac{2U}{b^2}$$

6. A $\quad \tau = \mu \frac{dv}{dy} = \mu \frac{r_1 \omega}{\delta} \qquad T = \tau 2\pi r_1 L \times r_1 \qquad \dot{W} = T\omega = 2\pi \frac{\mu r_1^3 \omega^2 L}{\delta}$

7. D $\quad V = \frac{0.0004}{\pi \times 0.06^2} = 0.0354 \text{ m/s} \qquad \text{Re} = \frac{0.0354 \times 0.12}{0.9 \times 10^{-6}} = 4720$

$\qquad \therefore \text{Use } n = 6.$

$$u_{\text{max}} = \frac{(n+1)(2n+1)}{2n^2} V = \frac{7 \times 13}{2 \times 6^2} \times 0.0354 = 0.045 \text{ m/s}$$

8. B $\quad \text{Re} = \frac{10 \times 0.4}{2 \times 10^{-4}} = 20\,000.$ From Table 7.1 select $n \cong 6.5.$ $f = \frac{1}{6.5^2} = 0.024$

$$\Delta p = \gamma h_L = \gamma f \frac{L}{D} \frac{V^2}{2g} = 9000 \times 0.024 \times \frac{100}{0.4} \times \frac{10^2}{2 \times 9.81} = 275\,000 \text{ Pa}$$

9. C $\quad V = \frac{Q}{A} = \frac{0.02}{\pi \times 0.04^2} = 3.98 \text{ m/s} \qquad \therefore \text{Re} = \frac{3.98 \times 0.08}{10^{-6}} = 3.18 \times 10^5$

$$\frac{e}{D} = \frac{0.15}{80} = 0.0019. \text{ Moody diagram: } f = 0.024$$

$$h_L = f \frac{L}{D} \frac{V^2}{2g} = 0.024 \times \frac{40}{0.08} \times \frac{3.98^2}{2 \times 9.81} = 9.7 \text{ m}$$

10. C $\quad h_L = \frac{\Delta p}{\gamma} = \frac{200\,000}{9810} = 20.4 \text{ m.}$ Use the alternate equation:

$$Q = -0.965 \sqrt{\frac{9.81 \times 0.1^5 \times 20.4}{100}} \ln\left[\frac{0.00015}{3.7 \times 0.1} + \left(\frac{3.17 \times (10^{-6})^2 \times 100}{9.81 \times 0.1^3 \times 20.4}\right)^{0.5}\right]$$

$$= 0.0333 \text{ m}^3/\text{s}$$

11. A   Assume a completely turbulent flow. Moody's diagram with $e/D = 0.15/40 = 0.00375$ gives $f = 0.027$. The energy equation:

$$20 = \frac{V^2}{2g} + 0.5\frac{V^2}{2g} + 0.027 \times \frac{100}{0.04}\frac{V^2}{2g} \qquad \therefore V = 2.01 \text{ m/s}$$

Check: $\text{Re} = \frac{2.01 \times 0.04}{10^{-6}} = 8 \times 10^4$. Moody diagram: $f = 0.028$. OK.

12. B $R = \frac{2 \times 0.8}{1.6 + 2} = 0.444$ m

$$Q = \frac{1}{n}AR^{2/3}S^{1/2} = \frac{1}{0.012} \times 2 \times 0.8 \times 0.444^{2/3} \times \sqrt{0.001} = 2.45 \text{ m}^3\text{/s}$$

QUIZ NO. 2

1. $V \cong \sqrt{2gh} = \sqrt{2 \times 9.81 \times 0.2} = 1.981$ m/s when the water leaves the fountain.

Assuming $T \cong 15°C$ $\text{Re} = \frac{1.981 \times 0.004}{1.14 \times 10^{-6}} = 6950$ $\therefore$ Turbulent

2. $\text{Re} = \frac{\omega r_1 \delta}{\nu} = \frac{(1000 \times 2\pi/60) \times 0.02 \times (0.022 - 0.02)}{1.8 \times 10^{-5}} = 233$ $\therefore$ Laminar

3. $V = \frac{Q}{A} = \frac{0.060/(90 \times 60)}{\pi \times 0.004^2} = 0.221$ m/s $\text{Re} = \frac{0.221 \times 0.008}{1.007 \times 10^{-6}} = 1760$

    $\therefore$ Laminar $L_E = 0.065 \times 1760 \times 0.008 = 0.913$ m so profile is parabolic.

4. $\tau_0 = \frac{r_0 \Delta p}{2L} = \frac{0.004 \times 1200}{2 \times 15} = 0.16$ Pa

$$\Delta p = \frac{8\mu LV}{r_0^2} = \frac{8 \times 0.000656 \times 15V}{0.004^2} = 1200 \qquad \therefore V = 0.244 \text{ m/s}$$

$$f = \frac{8\tau_0}{\rho V^2} = \frac{8 \times 0.16}{992 \times 0.244^2} = 0.0217$$

5. $Q = \int u(y)dy = \int_0^b \left(\frac{1}{2\mu}\frac{dp}{dx}(y^2 - by) + \frac{U}{b}y \right)dy = \frac{1}{2\mu}\frac{dp}{dx}(b^3/3 - b^3/2) + \frac{Ub}{2}$

$$0 = -\frac{1}{2\mu}\frac{dp}{dx}\frac{b^3}{6} + \frac{Ub}{2} \qquad \therefore \frac{dp}{dx} = \frac{6\mu U}{b^2}$$

6. $u(y) = \dfrac{1}{2\mu} \dfrac{\gamma \Delta h}{L}(by - y^2) = \dfrac{1}{2 \times 0.001} \times 9810 \times 0.0002(0.01y - y^2)$

   $= 980(0.01y - y^2)$

   $Q = \displaystyle\int_0^{0.005} 981(0.01y - y^2)80 dy = 981 \times 80 \times \left(0.01\dfrac{0.005^2}{2} - \dfrac{0.005^3}{3} \right)$

   $= 0.00653 \text{ m}^3/\text{s}$

7. $T = \tau 2\pi r_1 L \times r_1 = 2\pi \dfrac{\mu r_1^3 \omega L}{\delta} = 2\pi \dfrac{0.1 \times 0.03^3 \times 40 \times 0.2}{0.0004} = 0.339 \text{ N} \cdot \text{m}$

8. $V = \dfrac{0.0004}{\pi \times 0.06^2} = 0.0354 \text{ m/s} \qquad \dfrac{e}{D} = \dfrac{0.00026}{0.12} = 0.0022$

   $\text{Re} = \dfrac{0.0354 \times 0.12}{0.9 \times 10^{-6}} = 4720$

   From the Moody diagram, this is very close to a smooth pipe so the pipe is "smooth."

9. $\text{Re} = \dfrac{10 \times 0.4}{2 \times 10^{-4}} = 20\ 000.$ Table 7.1 gives $n \cong 6.5.$ $f = \dfrac{1}{6.5^2} = 0.024$

   $\tau_0 = \dfrac{1}{8}\rho V^2 f = \dfrac{1}{8} \times 917 \times 10^2 \times 0.024 = 275 \text{ Pa}$

   $u_\tau = \sqrt{\dfrac{\tau_0}{\rho}} = \sqrt{\dfrac{275}{917}} = 0.548 \text{ m/s} \qquad \therefore \delta_v = \dfrac{5v}{u_\tau} = \dfrac{5 \times 2 \times 10^{-4}}{0.548} = 0.00182 \text{ m}$

10. $V = \dfrac{Q}{A} = \dfrac{0.02}{\pi \times 0.04^2} = 3.98 \text{ m/s} \qquad \therefore \text{Re} = \dfrac{3.98 \times 0.08}{9 \times 10^{-6}} = 3.18 \times 10^4$

   Moody diagram: $\dfrac{e}{D} = \dfrac{0.15}{80} = 0.0019$ so that $f = 0.0375$

   $h_L = f\dfrac{L}{D}\dfrac{V^2}{2g} = 0.028 \times \dfrac{40}{0.08} \times \dfrac{3.98^2}{2 \times 9.81} = 11.3 \text{ m}$

11. $R = \dfrac{A}{P} = \dfrac{0.02 \times 0.06}{0.02 + 0.02 + 0.06 + 0.06} = 0.0075 \text{ m}$

   $D = 4R = 4 \times 0.0075 = 0.03 \text{ m}$

$$Q = -0.965 \sqrt{\frac{9.8 \times 0.03^5 \times 0.612}{20}} \ln\left(\frac{3.17 \times 0.804^2 \times 10^{-12} \times 20}{9.8 \times 0.03^3 \times 0.612}\right)^{.5}$$

$$= 0.00063 \text{ m}^3/\text{s}$$

12. Assume a completely turbulent flow. Moody's diagram with $e/D = 0.15/80 = 0.0019$ gives $f = 0.023$. The energy equation is then, using

$$V = \frac{Q}{A} = \frac{0.04}{\pi \times 0.08^2} = 7.96 \text{ m/s}$$

$$H_P = 0.5\frac{7.96^2}{2g} + 1.0\frac{7.96^2}{2g} + 0.023 \times \frac{200}{0.08}\frac{7.96^2}{2g} + 40 \qquad \therefore H_P = 231 \text{ m}$$

$$\therefore \dot{W}_P = \frac{\dot{m}gH_P}{\eta_P} = \frac{(1000 \times 0.04) \times 9.8 \times 231}{0.88} = 103\,000 \text{ W} \quad [\text{See Eq. (4.25)}]$$

13. $R = \dfrac{2 \times 1.2}{2.4 + 2} = 0.545 \text{ m}$

$$Q = \frac{1}{n} AR^{2/3} S^{1/2} = \frac{1}{0.012} \times 2 \times 1.2 \times 0.545^{2/3} \times \sqrt{0.001} = 4.22 \text{ m}^3/\text{s}$$

Chapter 8

QUIZ NO. 1

1. A $Re = \dfrac{VD}{\nu} = \dfrac{35 \times 0.041}{1.42 \times 10^{-5}} = 101\,000 \qquad \therefore C_D \cong 0.25$ from Fig. 8.2 (rough)

2. B $C_D = \dfrac{F_D}{\frac{1}{2}\rho V^2 A}$ and $p = \frac{1}{2}\rho V^2$ $C_D = \dfrac{pA}{\frac{1}{2}\rho V^2 A} = \dfrac{\frac{1}{2}\rho V^2 \times A}{\frac{1}{2}\rho V^2 A} = 1.0$

3. D $F_D = \dfrac{1}{2}\rho V^2 A C_D = \dfrac{1}{2} \times 1.2 \times 10^2 \times 0.1 \times 0.2 \times 1.1 = 1.32 \text{ N}$

4. D Assume $C_D = 0.2$ (see Fig. 8.2 with $Re > 4 \times 10^5$):

$$F_D = \frac{1}{2}\rho V^2 A C_D \qquad 4 = \frac{1}{2} \times 1.2 \times V^2 \times \pi \times 0.1^2 \times 0.2 \qquad \therefore V = 32.6 \text{ m/s}$$

Check: $Re = \dfrac{32.6 \times 0.20}{1.81 \times 10^{-5}} = 3.6 \times 10^5$ Good.

5. C $F_L = \frac{1}{2}\rho V^2 A C_L = \frac{1}{2} \times 1000 \times 15^2 \times (2.2 \times 0.5) \times 0.8 = 99\,000$ N

6. A Read $C_L \cong 1.0$ at $8°$ from Fig. 8.6.

$$F_L = W = 120\,000 = \frac{1}{2} \times 1.164 \times V^2 \times 20 \times 1.0 \qquad \therefore V = 102 \text{ m/s}$$

7. C $-\dfrac{\partial \psi}{\partial x} = 10 \qquad \therefore \psi = -10x + f(y) \qquad \dfrac{\partial \psi}{\partial y} = u = 0 = \dfrac{df}{dy}$

$\therefore f = C$ and $\psi = -10x$ if we let $C = 0$.

8. D $u = \dfrac{\partial \psi}{\partial y} = \dfrac{20y}{x^2 + y^2} \qquad v = -\dfrac{\partial \psi}{\partial x} = -\dfrac{20x}{x^2 + y^2}$

Along x-axis, $u = 0$, $v = -20/x$.

$$\frac{V^2}{2} + \frac{p}{\rho} = \frac{V_\infty^2}{2} + \frac{p_\infty}{\rho} \qquad \therefore p = -\frac{1.2}{2} \times \left(\frac{20}{x}\right)^2 = -\frac{240}{x^2}$$

9. B $\psi = 10r\sin\theta - \dfrac{40\sin\theta}{r} \qquad r_c = \sqrt{\dfrac{\mu}{U_\infty}} = \sqrt{\dfrac{40}{10}} = 2$ m

$$v_\theta = -\frac{\partial \psi}{\partial r}\bigg|_{r=r_c} = -\left(10 + \frac{40}{r_c^2}\right)\sin\theta = 20\sin\theta$$

10. D $5 \times 10^5 = \dfrac{U_\infty x_T}{\nu} = \dfrac{10 x_T}{1.81 \times 10^{-5}} \qquad \therefore x_T = 0.905$ m

11. A $\tau_0 = 0.14\rho U_\infty^2 \dfrac{d\delta}{dx} = 0.35\rho U_\infty^2 \sqrt{\dfrac{\nu}{xU_\infty}} \qquad \text{Drag} = \displaystyle\int_0^L \tau_0 w\,dx =$

$$= 0.7\rho U_\infty^2 \sqrt{\nu/U_\infty} \int_0^4 x^{-1/2}dx = 2.8 \times 1.2 \times 1^2 \sqrt{\frac{1.51 \times 10^{-5}}{1}} = 0.013 \text{ N}$$

12. C Use Eq. (8.82):

$$v_{\max} = \frac{1}{2}\sqrt{\frac{\nu U_\infty}{x}}(\eta F' - F)_{\max} = \sqrt{\frac{10^{-6} \times 0.8}{3}} \times 0.8605 = 4.4 \times 10^{-4} \text{ m/s}$$

QUIZ NO. 2

1. $Re = \dfrac{VD}{v} = \dfrac{2 \times 0.10}{1.33 \times 10^{-5}} = 15\,040.$ $L/D = 4/0.1 = 40$; refer to Table 8.1 (one end is assumed fixed and one end free so use $L/D = 80$) and Fig. 8.2: assume a factor of 0.95 so that $C_D = 0.95 \times 1.2 = 1.14.$

2. $Re = 10 \times 0.05/1.51 \times 10^{-5} = 33\,000$ $\quad \therefore C_D = 1.2$ from Fig. 8.2.

   $F_D = \dfrac{1}{2}\rho V^2 A C_D = \dfrac{1}{2} \times 1.2 \times 10^2 \times 0.1 \times 0.8 \times 1.25 \times 0.67 = 4.0$ N where the

   factor 0.67 is from Table 8.1.

3. $F_D = \dfrac{1}{2}\rho V^2 A C_D = \dfrac{1}{2} \times 1.25 \times 50^2 \times 2.2 \times 2.2 \times 1.1 = 8300$ N

   $8300 \times (3 + 1.1) = 34\,000$ N·m. (The force on the post is negligible.)

4. Assume $St = 0.21.$ $0.21 = \dfrac{fD}{V}$ $\quad \therefore V = \dfrac{0.16 \times 0.04}{0.21} = 0.0305$ m/s

   $Re = \dfrac{0.0305 \times 0.04}{1.81 \times 10^{-5}} = 67.4.$ Too low. Assume $St = 0.16.$

   $0.16 = \dfrac{fD}{V}$ $\quad \therefore V = \dfrac{0.16 \times 0.04}{0.16} = 0.04$ m/s

   $Re = \dfrac{0.04 \times 0.04}{1.81 \times 10^{-5}} = 88$ $\quad$ OK.

5. $W = 2000 \times 9.81 + 4000 = 23\,600$ N. The stall speed occurs with $C_L \cong 1.7$:

   $F_L = 23\,600 = \dfrac{1}{2} \times 1.2 \times V^2 \times 25 \times 1.7$ $\quad \therefore V = 33.3$ m/s

6. $\dfrac{\partial \phi}{\partial x} = u = 20x$ $\quad \therefore \phi = 10x^2 + f(y)$

   $\dfrac{\partial \phi}{\partial y} = v = 20y = \dfrac{df}{dy}$ $\quad \therefore f = 10y^2 + C$ $\quad$ and $\quad \phi = 10(x^2 + y^2)$

   (We let $C = 0$.)

7. $u = \dfrac{\partial \psi}{\partial y} = \dfrac{20y}{x^2 + y^2}$ $v = -\dfrac{\partial \psi}{\partial x} = -\dfrac{20x}{x^2 + y^2}$ $a_x = u\dfrac{\partial u}{\partial x} + v\dfrac{\partial u}{\partial y} + \dfrac{\partial u}{\partial t}$

$$= -\dfrac{20x}{x^2 + y^2} \times \dfrac{(x^2 + y^2)\times 20 - 20y(2y)}{(x^2 + y^2)^2} = -\dfrac{20}{4}\times\dfrac{20}{4^2} = -6.25 \text{ m/s}^2$$

8. $v_r = \dfrac{\partial \phi}{\partial r} = 10\cos\theta + \dfrac{40}{r}$ $v_\theta = -\dfrac{1}{r}\dfrac{\partial \phi}{\partial \theta} = -10\sin\theta$. If $v_r = 0$ and $v_\theta = 0$, then

$\theta = 180°$ and $r = 4$ m which in rectangular coordinates is $(4,0)$, only one point.

9. $\psi = 10y + \dfrac{10\pi}{2\pi}\theta = 10r\sin\theta + 5\theta$ $v_r = 10\cos\theta + \dfrac{5}{r}$ $v_\theta = -10\sin\theta$

Stagnation point: $\theta = \pi$ and $r = 0.5$ so that $\psi = 5\pi$.

At $\theta = \pi/2$ $\psi = 5\pi = 10y + 5\times\dfrac{\pi}{2}$ $\therefore y = \pi/4$ m

10. $u = U_\infty\dfrac{y}{\delta}$ $\tau_0 = \rho\dfrac{d}{dx}\displaystyle\int_0^\delta \dfrac{U_\infty y}{\delta}\left(U_\infty - \dfrac{U_\infty y}{\delta}\right)dy = \dfrac{1}{6}\rho U_\infty^2\dfrac{d\delta}{dx}$

$\tau_0 = \mu\left.\dfrac{\partial u}{\partial y}\right|_{y=0} = \mu\dfrac{U_\infty}{\delta} = \dfrac{1}{6}\rho U_\infty^2\dfrac{d\delta}{dx}$ $\therefore \delta\dfrac{d\delta}{dx} = \dfrac{6\mu}{\rho U_\infty}$

or $\delta^2 = 12\dfrac{\nu x}{U_\infty}$ $\therefore \delta = 3.46\sqrt{\dfrac{\nu x}{U_\infty}}$

11. It should be displaced outward $4\delta_d$ so continuity is satisfied. If the boundary layer is turbulent it would be [see Eq. (8.69)]:

$$\delta_d = 4\times 0.048x\left(\dfrac{\nu}{U_\infty x}\right)^{1/5} = 0.192x\left(\dfrac{1.51\times 10^{-5}}{10x}\right)^{1/5} = 0.0132x^{0.8}$$

12. $\text{Re} = \dfrac{U_\infty L}{\nu} = \dfrac{16\times 3}{1.8\times 10^{-5}} = 2.7\times 10^6$ $\text{Drag} = \dfrac{1}{2}\rho U_\infty^2 Lw\times 0.073\left(\dfrac{\nu}{U_\infty L}\right)^{1/5}$

$$= \dfrac{0.073}{2}\times 1.2\times 16^2\times 6\times\left(\dfrac{1.8\times 10^{-5}}{16\times 3}\right)^{0.2} = 3.49 \text{ N}$$

13. $F_D \times U_\infty = \frac{1}{2}\rho U_\infty^3 A C_f$

$$= \frac{1}{2} \times 1.2 \times 12^3 \times \pi \times 150 \times 1000 \times 0.073 \left(\frac{1.8 \times 10^{-5}}{12 \times 1000}\right)^{0.2} = 596\,000 \text{ W}$$

Chapter 9

QUIZ NO. 1

1. C Assume the speed of a small wave in water is 1450 m/s. The distance is
   $d = V\Delta t = 1450 \times 0.45 = 652$ m.

2. B $M = \dfrac{V}{c} = \dfrac{200}{\sqrt{1.4 \times 287 \times 223}} = 0.668$

3. D $\sin\alpha = \dfrac{1}{1.68}$ so $\tan\alpha = 0.7408$

   $\therefore L = \dfrac{200}{\tan\alpha} = 270$ m $d = \sqrt{270^2 + 200^2} = 336$ m

4. A Check: $p_e = 0.5283 p_0 = 0.5283 \times 250 = 132.1$ kPa $\therefore M_e < 1$

   $$\frac{250}{150} = \left(1 + \frac{1.4-1}{2}M^2\right)^{3.5}   \therefore M = 0.886$$

   $$\dot{m} = 250\,000 \times 0.886 \times (10 \times 10^{-4})\sqrt{\frac{1.4}{287 \times 293}}\left(1 + \frac{0.4}{2} \times 0.886^2\right)^{-3}$$

   $$= 0.584 \text{ kg/s}$$

5. A $\dfrac{A}{A^*} = \dfrac{24^2}{10^2} = 5.76.$ From Table D.1 $\dfrac{p_e}{p_0} = 0.01698$ $\therefore p_e = 6.79$ kPa

6. C $M_1 = \dfrac{V_1}{c_1} = \dfrac{600}{\sqrt{1.4 \times 287 \times 291}} = 1.76$ so that $M_2 = 0.626$ from Table D.2.

   $T_2 = 1.502 \times 291 = 437$ K   and

   $V_2 = 0.626 \times \sqrt{1.4 \times 287 \times 437} = 262$ m/s

7. B Table D.2 provides $M_2 = 0.577$. $T_2 = 1.688 \times 288 = 486$ K

$$V_2 = 0.5774 \times \sqrt{1.4 \times 287 \times 486} = 255 \text{ m/s}$$

$$V_1 = 2 \times \sqrt{1.4 \times 287 \times 288} = 680 \text{ m/s}$$

$$V_{\text{induced}} = V_1 - V_2 = 680 - 255 = 425 \text{ m/s}$$

8. D At $\dfrac{A_1}{A^*} = \dfrac{20^2}{10^2} = 4.0$ in Table D.1: $M_1 = 2.94$, $p_1 = 400 \times .0299 = 12$ kPa

Table D.2: At $M_1 = 2.94$: $M_2 = 0.479$ $p_2 = 9.92 \times 12.0 = 119$ kPa

9. C $M_1 = \dfrac{900}{\sqrt{1.4 \times 287 \times 313}} = 2.54$. From Fig. 9.11 $\beta = 42°$.

$M_{1n} = 2.54 \sin 42° = 1.70$ $M_{2n} = 0.64$ and

$p_2 = 3.2 \times 60 = 192$ kPa abs

10. A Table D.3 at $M_1 = 2.4$: $\theta_1 + \theta_2 = 36.75 + 40 = 76.75°$ $\therefore M_2 = 4.98$

QUIZ NO. 2

1. The wave travels at a constant speed, so the time for the wave to reach the object is 0.23 s; the distance is $d = V \Delta t = 1450 \times 0.23 = 333$ m.

2. The speed of sound in air is approximately 340 m/s so the distance is about $d = V \Delta t = 340 \times 1.5 = 510$ m.

3. $\sin \alpha = \dfrac{1}{3.49}$ so $\tan \alpha = 0.299$

$$\therefore L = \dfrac{200}{\tan \alpha} = 669 \text{ m} \qquad d = \sqrt{669^2 + 200^2} = 672 \text{ m}$$

4. use $c = 340$ m/s, $\rho = 1.22$ kg/m³.

$$c^2 = \dfrac{\Delta p}{\Delta \rho} \qquad \therefore 340^2 = \dfrac{5}{\Delta \rho} \qquad \therefore \Delta \rho = \dfrac{5}{340^2} = 4.33 \times 10^{-5} \text{ kg/m}^3$$

Ideal-gas law: $\Delta p = \rho R \Delta T + \Delta \rho R T$

$5 = 1.21 \times 287 \times \Delta T + 4.33 \times 10^{-5} \times 287 \times 288 \qquad \therefore \Delta T = 0.00409°C$

Energy: $\dfrac{V^2}{2} + c_p T = \text{const}$ so $V \Delta V = -c_p \Delta T$ where $V = c$

$\Delta V = -\dfrac{c_p \Delta T}{c} = -\dfrac{1.0 \times 0.00409}{340} = -1.20 \times 10^{-5}$ m/s

5. $p_e = 0.5283 p_0 = 0.5283 \times 250 = 132.1$ kPa $\qquad \therefore M_e = 1 \quad (M_e > M_r)$

Use Eq. (9.33): $\dot{m} = 250\,000 \times (10 \times 10^{-4}) \sqrt{\dfrac{1.4}{287 \times 293}} \times 1.2^{-3} = 0.590$ kg/s

6. $\dfrac{A}{A^*} = 1.688 \qquad$ or $\qquad \dfrac{d^2}{(d^*)^2} = 1.688 \qquad \therefore d = \sqrt{1.688 \times 12^2} = 15.59$ cm

7. From Table D.1, $M_1 = 2.64$ so that $V_1 = 2.64 \times \sqrt{1.4 \times 287 \times 293} = 906$ m/s

$T_2 = 2.280 \times 293 = 668$ K $\quad$ and $\quad V_2 = 0.5 \times \sqrt{1.4 \times 287 \times 668} = 259$ m/s

8. Table D.1: $\dfrac{A_1}{A^*} = \dfrac{16^2}{10^2} = 2.56 \qquad \therefore M_1 = 2.47 \qquad p_1 = 0.0613 \times 400 = 24.5$

Table D.2: $M_2 = 0.516 \qquad p_2 = 6.95 \times 24.5 = 170.3$ kPa

and $\qquad p_{02} = 0.511 \times 400 = 204$ kPa

Table D.1: $\dfrac{A_2}{A_e} = \dfrac{A_2}{A^*} \times \dfrac{A^*}{A_e} \qquad \dfrac{8^2}{10^2} = 1.314 \times \dfrac{A^*}{A_e} \qquad \therefore \dfrac{A_e}{A^*} = 2.05$

$p_e = 0.940 \times 204 = 192$ kPa

9. $M_1 = \dfrac{900}{\sqrt{1.4 \times 287 \times 313}} = 2.54.$ From Fig. 9.11, $\beta = 81°$

$\therefore M_{1n} = 2.54 \sin 81° = 2.51 \qquad \therefore M_{2n} = 0.512$

$p_2 = 7.18 \times 60 = 431$ kPa abs

$0.512 = M_2 \sin(81° - 20°) \qquad \therefore M_2 = 0.585$

$T_2 = 2.147 \times 313 = 672$ K $\qquad \therefore V_2 = 0.585\sqrt{1.4 \times 287 \times 672} = 304$ m/s

10. Table D.3 at $M_1 = 1.8$: $\theta = 20.72°$ $\theta_2 = 60.09 - 20.72 = 39.4°$

$$T_2 = T_1 \frac{T_0}{T_1} \frac{T_2}{T_0} = 293 \times \frac{1}{0.6068} \times 0.2784 = 134 \text{ K}$$

$$V_2 = 3.6\sqrt{1.4 \times 287 \times 134} = 835 \text{ m/s}$$

Final Exam No. 1

1. C $F = ma$ $F = M\dfrac{L}{T^2}$ $\therefore M = FT^2/L$

2. A $2\sigma L > \pi r^2 L \rho_s$ or $\sigma > \pi r^2 L \rho_s / 2$

3. A $p_w + \gamma_w h = \gamma_{Hg} H$ $40\,000 + 9810 \times 0.20 = 13.6 \times 9810 H$
   $\therefore H = 0.315$ m

4. D F_H and F_V can be positioned at the center of the circular arc (F_H produces no moment):

   $$P = F_V = \gamma \mathcal{V} = 9810 \times [(0.6 \times 1.2) + \pi \times 1.2^2/4] \times 1.4 = 18\,200 \text{ N}$$

5. D $\mathbf{a} = u\dfrac{\partial \mathbf{V}}{\partial x} + v\dfrac{\partial \mathbf{V}}{\partial y} + w\dfrac{\partial \mathbf{V}}{\partial z} + \dfrac{\partial \mathbf{V}}{\partial t}$

   $= 2xy(2y\mathbf{i} + t\mathbf{j}) + xt(2x\mathbf{i}) + x\mathbf{j} = -4(-2\mathbf{i} + 2\mathbf{i}) + 4(4\mathbf{i}) + 2\mathbf{j} = 16\mathbf{i} + 2\mathbf{j}$

6. C $\dfrac{DT}{Dt} = u\dfrac{\partial T}{\partial x} + v\dfrac{\partial T}{\partial y} + w\dfrac{\partial T}{\partial z} + \dfrac{\partial T}{\partial t} = 2xy(-0.2y) + yt(-0.2x) = -6.4°\text{C/s}$

7. D

8. A $\rho Q_1 = \dot{m}_2 + \dot{m}_3$ $1000 \times 0.01 = 1000 \times \pi \times 0.01^2 \times 20 + \dot{m}_3$
   $\dot{m}_3 = 3.72$ kg/s

9. B Assume incompressible flow:

   $(0.1w) \times 40 = 2w \times \bar{v} + 0.05w \times 20 + 0.05w \times 40$ $\therefore \bar{v} = 0.5$ m/s

10. A $\dot{W}_T = H_T \dot{m} g \eta_T = (25 - 5) \times 0.01 \times 1000 \times 9.81 \times 1 = 1962$ W (let $\eta_T = 1$)

11. C $\quad \dot{W}_p = \dot{m}\dfrac{p_2 - p_1}{\rho\eta_p} = 1000 \times \pi \times 0.03^2 \times 4 \times \dfrac{2000 - 120}{1000 \times 0.88} = 24.2$ kW $\qquad$ or

$\qquad$ 32.4 hp

12. A $\quad V_1 = V_2/5^2 = 2$ m/s $\qquad \dfrac{2^2}{2} + \dfrac{p_1}{1000} = \dfrac{50^2}{2} + \dfrac{\cancel{p_2}}{\cancel{\rho}} \qquad \therefore p_1 = 1\,248\,000$ Pa

$\qquad -F_x + 1\,248\,000 \times \pi \times 0.05^2 = 1000 \times \pi \times 0.05^2 \times 2(50 - 2)$

$\qquad \therefore F_x = 9050$ N

13. B $\quad -F_x = \dot{m}(V_{2x} - V_{1x}) = 1000 \times \pi \times 0.01^2 \times 40(40\cos 60° - 40) = -251$ N

14. D $\quad \rho V R/\mu$

15. D $\quad \dfrac{V_m^2}{l_m g_m} = \dfrac{V_p^2}{l_p g_p} \qquad \dfrac{V_p}{V_m} = \sqrt{\dfrac{l_p}{l_m}} = \sqrt{25} = 5$

$\qquad \dfrac{F_p}{F_m} = \dfrac{\rho_p V_p^2 l_p^2}{\rho_m V_m^2 l_m^2} = 1 \times 5^2 \times 25^2 \qquad \therefore F_p = 4 \times 15\,625 = 62\,500$ N

16. A $\quad \tau = 0.1 \times \dfrac{0.05 \times 800 \times 2\pi/60}{0.001} = 419$ N/m^2

$\qquad T = \tau AR = 419 \times (\pi \times 0.1 \times 0.25) \times 0.05 = 1.64$ N·m

17. C $\quad V = \dfrac{0.02}{\pi \times .02^2} = 15.9$ m/s $\qquad \therefore \text{Re} \cong 6 \times 10^5 \qquad \dfrac{e}{D} = \dfrac{.26}{40} = .0065$

$\qquad \therefore f \cong .033$

$\qquad \dfrac{0.85\dot{W}_p}{1000 \times 0.02 \times 9.81} = \left(.5 + .033 \times \dfrac{40}{.04} + 1\right)\dfrac{V^2}{2g} + 20 \qquad \therefore \dot{W}_p = 87\,200$ W

18. B $\quad 2 = \dfrac{1}{0.012} \times 2y \times \left(\dfrac{2y}{2 + 2y}\right)^{2/3} \times 0.002^{1/2} \qquad \therefore y \cong 0.55$ m

19. C $\quad \dot{W} = \dfrac{1}{2}\rho A V^3 C_D$

$\qquad \therefore \%\text{Savings} = \dfrac{C_{D1} - C_{D2}}{C_{D1}} \times 100 = \dfrac{0.96 - 0.7}{0.96} \times 100 = 27\%$

20. A $\quad r_c = \sqrt{\dfrac{\mu}{U_\infty}} = \sqrt{\dfrac{2}{2}} = 1 \text{ m}$

21. B $\quad \delta = 5\sqrt{\dfrac{vx}{U_\infty}} = 5\sqrt{\dfrac{0.804 \times 10^{-6} \times 2}{8}} = 0.00112 \text{ m}$

22. D $\quad \delta = 0.38x\left(\dfrac{v}{U_\infty x}\right)^{1/5} = 0.38 \times 2 \times \left(\dfrac{0.804 \times 10^{-6}}{8 \times 2}\right)^{0.2} = 0.026 \text{ m}$

23. A $\quad c = \sqrt{kRT} = \sqrt{1.4 \times 287 \times 288} = 340 \text{ m/s} \qquad M = \dfrac{600}{340} = 1.76$

24. C $\quad p_e = 0.5283 \times 200 \qquad \therefore p_e = 106 \text{ kPa} \ \text{(Use Table D.1)}$

25. B $\quad M_1 = \dfrac{400}{\sqrt{1.4 \times 287 \times 298}} = 1.16 \qquad \therefore M_2 = 0.868 \quad \text{(Use Table D.2)}$

$\qquad T_2 = 1.103 \times 298 = 329 \text{ K} \qquad \therefore V_2 = 0.868 \times \sqrt{1.4 \times 287 \times 329} = 316 \text{ m/s}$

26. D $\quad$ Table D.3: $\theta_1 = 36.75° \qquad 36.75° + 40° = 76.75° \qquad \therefore M_2 = 4.98$

Final Exam No. 2

1. $\dfrac{\text{m}^3}{\text{s}} = [k] \cdot \text{m}^2 \cdot \text{m}^{2/3} \cdot 1 \qquad \therefore [k] = \text{m}^{1/3}/\text{s}$

2. $T = \tau Ar = \mu\dfrac{\Delta u}{\Delta r} 2\pi r L \times r \qquad 0.05 = \mu\dfrac{2\pi \times 150/60}{0.05 - 0.048} \times 2\pi \times 0.05^2 \times 0.15$

$\qquad \therefore \mu = 0.0027 \text{ N} \cdot \text{s/m}$

3. $1.2P = 0.333 \times (9810 \times 0.5 \times 1.0 \times 2.4) \qquad \therefore P = 3270 \text{ N}$

4. $p_1 - p_A = \dfrac{\rho\Omega^2}{2}(r_1^2 - r_A^2) = \dfrac{1000 \times (150 \times 2\pi/60)^2}{2} \times 0.4^2$

$\qquad \therefore p_A = -19\,740 \text{ Pa}$

5. $\mathbf{V} = 2xy\mathbf{i} + xt\mathbf{j} = 2 \times 2 \times (-1)\mathbf{i} + 2 \times 2\mathbf{j} = -4\mathbf{i} + 4\mathbf{j}$

$$(n_x\mathbf{i}+n_y\mathbf{j})\cdot(-4\mathbf{i}+4\mathbf{j})=0 \qquad \therefore n_x = n_y \qquad n_x^2 + n_y^2 = 1 \qquad \therefore n_x = n_y = \frac{1}{\sqrt{2}}$$

$$\mathbf{n} = (\mathbf{i}+\mathbf{j})/\sqrt{2}$$

6. $$\frac{D\rho}{Dt} = u\frac{\partial\rho}{\partial x} + v\frac{\partial\rho}{\partial y} + w\frac{\partial\rho}{\partial z} + \frac{\partial\rho}{\partial t} = u\frac{\partial\rho}{\partial x} + w\frac{\partial\rho}{\partial z}$$

7. Use the density at sea level: $p = \frac{1}{2}\rho V^2 \cong \frac{1}{2}\times 1.2\times(180\ 000/3600)^2 = 1500$ Pa

$$F_{max} = pA = 1500\times 1\times 2 = 3000\ \text{N}$$

8. $\dot{m}_1 = \dot{m}_2$
$$\frac{140}{0.287\times 293}\times\frac{40}{60} = \frac{250}{0.287\times 298}\times\pi\times 0.2^2\times V_2$$
$$\therefore V_2 = 3.02\ \text{kg/s}$$

9. $$\dot{m} = \rho_2 A_2 V_2 = \frac{100}{0.287\times 293}\times\pi\times 0.02^2\times 30 = 0.0448\ \text{kg/s}$$

$$-\frac{\dot{W}_T}{\dot{m}} = \frac{V_2^2 - V_1^2}{2} + \frac{p_2}{\rho_2} - \frac{p_1}{\rho_1} \qquad -\frac{\dot{W}_T}{0.0448} = \frac{30^2}{2} + 287\times 293 - 287\times 473$$

$$\therefore \dot{W}_T = 2290\ \text{W} \qquad \text{(We used } p/\rho = RT)$$

10. $V_2 = V_1\dfrac{y_1}{y_2} = 20\times\dfrac{0.8}{y_2} = \dfrac{16}{y_2}$ $\qquad F_1 - F_2 = \dot{m}(V_2 - V_1)$

$$9810\times 0.4\times 0.8w - 9810\times\frac{y_2}{2}\times y_2 w = 1000\times 0.8w\times 20\times\left(\frac{16}{y_2} - 20\right)$$

$$0.64 - y_2^2 = 65\times\left(\frac{0.8 - y_2}{y_2}\right) \qquad \text{or} \qquad 0.8y_2 + y_2^2 = 65 \qquad \therefore y_2 = 7.65\ \text{m}$$

11. $\left.\begin{array}{l} 60\sin 30° = V_{r1}\sin\alpha_1 \\ 60\cos 30° = 20 + V_{r1}\cos\alpha_1 \end{array}\right\}$ $\qquad \therefore V_{r1} = 43.83\ \text{m/s} \qquad \alpha_1 = 43.19°$

Additional calculations follow:

$$\left.\begin{array}{l} V_2 \sin \beta_2 = 43.83 \sin 45° \\ V_2 \cos \beta_2 = 20 - 43.83 \cos 45° \end{array}\right\} \quad \therefore V_2 = 32.9 \text{ m/s} \qquad \beta_2 = 109.5°$$

$$-F_x = 1000 \times \pi \times 0.01^2 \times 60 \times (32.9 \cos 109.5° - 60 \cos 30°) = -986 \text{ N}$$

12. $T = f(D, \delta, \rho, \mu, \Omega)$

$$[T] = \frac{ML^2}{T^2}, \quad [D] = L, \quad [\delta] = L, \quad [\rho] = \frac{M}{L^3}, \quad [\mu] = \frac{M}{LT}, \quad [\Omega] = \frac{1}{T}$$

$$\frac{T}{\rho \Omega^2 D^5} = \left(\frac{D}{\delta}, \frac{\rho \Omega D^2}{\mu}\right)$$

13. Viscous effects would dominate so the Reynolds number is used:

$$\frac{V_m D_m}{V_m} = \frac{V_p D_p}{V_p} \qquad \therefore V_m = 0.006 \times 200 = 1.2 \text{ m/s}$$

14. $\dfrac{e}{D} = \dfrac{0.046}{60} = 0.00077 \qquad \therefore f \cong 0.019 \text{ (assume completely turbulent)}$

$$10 = \left(0.8 + 0.019 \times \frac{40}{0.06} + 1\right) \frac{V^2}{2g} \qquad \therefore V = 3.68 \text{ m/s}$$

$$\text{Re} = \frac{3.68 \times 0.06}{1.14 \times 10^{-6}} = 1.9 \times 10^5$$

Use $f = 0.02$ so $V = 3.60$ m/s and $\dot{m} = 1000 \times \pi \times 0.03^2 \times 3.6 = 10.2$ kg/s

15. $\psi = 5r \sin \theta - \dfrac{20 \sin \theta}{r} \qquad r_c = \sqrt{\dfrac{\mu}{U_\infty}} = \sqrt{\dfrac{20}{5}} = 2$ m

$$v_\theta = -\left.\frac{\partial \psi}{\partial r}\right|_{r=r_c} = -\left(5 + \frac{20}{2^2}\right) \sin \theta = -10 \sin \theta$$

$$p_{min} = p_0 - \rho \frac{v_\theta^2}{2} = 100\,000 - 1000 \frac{100 \times 1^2}{2} = 50\,000 \text{ Pa}$$

16. $\text{Drag} = \frac{1}{2}\rho U_\infty^2 Lw \times 0.073 \left(\dfrac{\nu}{U_\infty L} \right)^{1/5}$

$$= \frac{0.073}{2} \times 1000 \times 5^2 \times 2 \times \left(\frac{10^{-6}}{5 \times 2} \right)^{0.2} = 73 \text{ N}$$

17. $p = \gamma h = 9810 \times 13.6 \times 0.05 = 6670 \text{ Pa}$

$$\rho_1 = \frac{p_1}{RT_1} = \frac{101}{0.287 \times 313} = 1.12 \text{ kg/m}^3$$

$$T_0 = T_1 \left(\frac{p_0}{p_1} \right)^{\frac{k-1}{k}} = 313 \times \left(\frac{101 + 6.67}{101} \right)^{0.2857} = 318.8 \text{ K}$$

$$c_p T_0 = \frac{V^2}{2} + c_p T_1 \qquad \therefore V = \sqrt{2 \times 1000 \times (318.8 - 313)} = 108 \text{ m/s}$$

18. $\text{M}_{1n} = 2 \sin 40° = 1.28 \qquad \therefore \text{M}_{2n} = 0.796 \qquad \text{From Fig. 9.11 } \theta = 11°$

$$\text{M}_2 = \frac{0.796}{\sin(40° - 11°)} = 1.64 \qquad \theta_2 = \theta_1 = 11° \qquad \therefore \beta_2 \cong 50°$$

$$\beta_3 = 50 - 11 = 39°$$

θ_2 is the angle through which the flow must turn to be parallel to the wall.

INDEX